Oberwolfach Seminars
Volume 35

Ilia Itenberg
Grigory Mikhalkin
Eugenii Shustin

Tropical
Algebraic
Geometry

Birkhäuser
Basel · Boston · Berlin

Authors:

Ilia Itenberg
IRMA, Université Louis Pasteur
7 rue René Descartes
67084 Strasbourg Cedex
France
e-mail: itenberg@math.u-strasbg.fr

Grigory Mikhalkin
Department of Mathematics
University of Toronto
Toronto, Ont. M5S 2E4
Canada
e-mail: mikha@math.toronto.edu

Eugenii Shustin
School of Mathematical Sciences
Raymond and Beverly Sackler Faculty of Exact Sciences
Tel Aviv University
Ramat Aviv, 69978 Tel Aviv
Israel
e-mail: shustin@post.tau.ac.il

2000 Mathematics Subject Classification 14M25, 14N35, 14N10, 52B20, 14P25, 14H99

Library of Congress Control Number: 2009923622

Bibliographic information published by Die Deutsche Bibliothek
Die Deutsche Bibliothek lists this publication in the Deutsche Nationalbibliografie;
detailed bibliographic data is available in the Internet at <http://dnb.ddb.de>.

ISBN 978-3-0346-0047-7 Birkhäuser Verlag, Basel – Boston – Berlin

First edition 2007

© 2009 Birkhäuser Verlag AG
Basel · Boston · Berlin
P.O. Box 133, CH-4010 Basel, Switzerland
Part of Springer Science+Business Media
Printed on acid-free paper produced from chlorine-free pulp. TCF ∞
Printed in Germany

ISBN 978-3-0346-0047-7   e-ISBN 978-3-0346-0048-4

9 8 7 6 5 4 3 2 1   www.birkhauser.ch

Contents

Preface

This book is based on the lectures given at the Oberwolfach Seminar on Tropical Algebraic Geometry in October 2004.

Tropical Geometry first appeared as a subject of its own in 2002, while its roots can be traced back at least to Bergman's work [1] on logarithmic limit sets. Tropical Geometry is now a rapidly developing area of mathematics. It is intertwined with algebraic and symplectic geometry, geometric combinatorics, integrable systems, and statistical physics. Tropical Geometry can be viewed as a sort of algebraic geometry with the underlying algebra based on the so-called tropical numbers. The tropical numbers (the term "tropical" comes from computer science and commemorates Brazil, in particular a contribution of the Brazilian school to the language recognition problem) are the real numbers enhanced with negative infinity and equipped with two arithmetic operations called tropical addition and tropical multiplication. The tropical addition is the operation of taking the maximum. The tropical multiplication is the conventional addition. These operations are commutative, associative and satisfy the distribution law. It turns out that such tropical algebra describes some meaningful geometric objects, namely, the Tropical Varieties. From the topological point of view the tropical varieties are piecewise-linear polyhedral complexes equipped with a particular geometric structure coming from tropical algebra. From the point of view of complex geometry this geometric structure is the worst possible degeneration of complex structure on a manifold. From the point of view of symplectic geometry the tropical variety is the result of the Lagrangian collapse of a symplectic manifold (along a singular fibration by Lagrangian tori).

The target audience of the Oberwolfach seminar was graduate students. The seminar was designed to introduce young mathematicians to this perspective research field, including presentation of basic notions and motivations for tropical algebraic geometry as well as demonstration of some of its striking applications. During the preparation of these lecture notes for publication, we adapted the notes to a wider audience, both beginners and established researchers. As a result, the discussions in this book are more detailed and contain some new results that were obtained after the seminar itself.

Besides a general introduction to tropical geometry, we discuss the concepts of complex and non-Archimedean amoebas, as well as the patchworking construction and enumerative tropical geometry. For a more advanced study of these topics, we recommend the articles [8, 22, 28, 39, 40, 41, 42, 48, 59].

We do not in this book attempt to cover all facets of tropical geometry. For instance, we do not discuss the combinatorial aspects of tropical varieties (see, for example, [31, 54, 62, 64]), or abstract tropical varieties of dimension greater than 1 [18, 34, 35]. Furthermore, we do not touch various other branches of tropical mathematics, but only recommend some references: [36, 64] (computational aspects), [5, 15, 53] (max-plus algebra), [9, 32, 37, 50, 52] (tropical mathematics in applied problems).

The book consists of three chapters. The first chapter, "Introduction to tropical geometry" by G. Mikhalkin, is a basic treatment of tropical varieties and their relation to classical geometry, in particular the theory of amoebae. Special emphasis is put on tropical curves. The second chapter, "Patchworking of algebraic varieties" by E. Shustin, deals with the patchworking construction in algebraic geometry, the link between real algebraic geometry and tropical geometry. The chapter starts with the original Viro method of gluing real algebraic hypersurfaces, then goes through various modifications and generalizations of the Viro method. In the final section the patchworking construction is used to prove Mikhalkin's correspondence theorem between real and complex algebraic curves on toric surfaces on one side and plane tropical curves on the other side. The third chapter, "Applications of tropical geometry to enumerative geometry" by I. Itenberg, presents various applications, based on Mikhalkin's correspondence theorem, of tropical geometry in real and complex enumerative geometry. These applications mostly concern calculation of Gromov–Witten invariants and Welschinger invariants (the latter invariants can be seen as real counterparts of genus zero Gromov–Witten invariants).

Each chapter is supplemented by exercises, most of which were proposed to and discussed by the participants of the seminar.

Acknowledgements. We are grateful to Mathematisches Forschungsinstitut Oberwolfach for a unique opportunity to run a seminar on tropical algebraic geometry.

Our special thanks go to Oliver Wienand. We are very grateful to him for taking notes of our lectures and helping in expanding them for publication. His role in the work on this book is hard to overestimate.

The first author was partially supported by the ANR-05-0053-01 grant of Agence Nationale de la Recherche and a grant of Université Louis Pasteur, Strasbourg. The second author is supported in part by NSERC. The third author was supported by the Hermann-Minkowski-Minerva center for Geometry at the Tel Aviv University and by the grant no. 465/04 from the Israel Science Foundation. The first and the third authors were partially supported by a grant of the Ministère des Affaires Etrangères, France and the Ministry of Science and Technology, Israel.

Preface to the second edition

We are happy to observe that Tropical Geometry has become an even more popular subject. A number of new directions for tropical methods has emerged and developed. As a result the collection of new tropical research papers is too large to make an exhaustive list.

In this edition we have corrected some of the misprints from the first edition and added the references [2], [23], [43], [65] to new lecture notes similar in spirit to those from this book.

Chapter 1

Introduction to tropical geometry

In this section the notion of an amoeba of a variety will be introduced and several examples of such amoebas are given. Then we consider a degenerations process where an amoeba becomes a piecewise-linear object.

1.1 Images under the logarithm

We start with an algebraic variety over $(\mathbb{C}^*)^n$. Namely, let I be an ideal in the ring of polynomials in n variables over $\mathbb{C}$. Then the variety is given by

$$V = \{x \in (\mathbb{C}^*)^n \mid f(x) = 0 \text{ for all } f \in I\}.$$

We define the map $\operatorname{Log} : (\mathbb{C}^*)^n \longrightarrow \mathbb{R}^n$,

$$(z_1, z_2, \ldots, z_n) \longmapsto (\log|z_1|, \log|z_2|, \ldots, \log|z_n|).$$

Definition 1.1. Let $V \subset (\mathbb{C}^*)^n$ be an algebraic variety. Then we define its *amoeba* as $A(V) = \operatorname{Log}(V)$. This is a subset of $\mathbb{R}^n$:

$$\mathcal{A}(V) = \operatorname{Log} V \subset \mathbb{R}^n \ .$$

0-dimensional amoebas

If V is 0-dimensional, then it is just a collection of points and so is $\operatorname{Log}(V)$.

Amoeba of a line in $\mathbb{P}^2$

For our first example of an amoeba of a 1-dimensional variety, consider the case when $V \subset (\mathbb{C}^*)^2 \subset \mathbb{CP}^2$ is a line given by equation

$$z + w + 1 = 0. \tag{1.1}$$

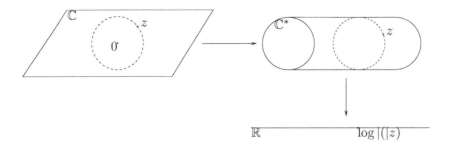

Figure 1.1: Going from $\mathbb{C}$ to $\mathbb{R}$ with Log.

Solving it for w we get $w = -z - 1$.

Set $x = \log|z|$, $y = \log|w|$. The value of x does not determine y as it also depends on the argument of z, but we get the following inequalities. If $x \geq 0$, then

$$\log(e^x - 1) \leq y \leq (1 + e^x);$$

if $x \leq 0$, then

$$\log(1 - e^x) \leq y \leq (1 + e^x).$$

Assume now that $V \subset (\mathbb{C}^*)^2$ is given by $az + bw + c = 0$ with $a, b, c \in (\mathbb{C}^*)$. In coordinates $z' = \frac{a}{c}z$, $w' = \frac{b}{c}w$, we get the equation (1.1) again. Thus the amoeba $\mathcal{A}(V)$ in this case is just a translation of the one pictured in Figure 1.2 by

$$x \mapsto x + \log|c| - \log|a|, \ y \mapsto y + \log|c| - \log|b|.$$

If $a, b, c \in R$, then the variety V is defined over the reals and thus we may consider its real locus $\mathbb{R}V$. Note that in this case the amoeba $\mathrm{Log}(\mathbb{R}V)$ is the boundary of the amoeba $\mathcal{A}(V)$.

In the case of a general hypersurface $V \subset (\mathbb{C}^*)^n$ defined over the reals, $\mathrm{Log}(\mathbb{R}V)$ is a subset of the discriminant locus of $\mathrm{Log}|_V : V \to \mathbb{R}^n$, i.e., the locus of the critical values of $\mathrm{Log}|_V$. There is a class of real varieties $\mathbb{R}V$ such that $\mathrm{Log}(\mathbb{R}V)$ *coincides* with the corresponding discriminant locus (see [38] for the case of curves). These varieties have some extremal topological and geometric properties. The lines (and hyperplanes in $(\mathbb{C}^*)^n$) are examples of such extremal hypersurfaces.

Geometric properties of the amoeba

Amoebas of hypersurfaces have the following properties, see [16], [11], [72], [38], [51], [44]. Let $V \subset (\mathbb{C}^*)^n$ be the zero locus of a polynomial

$$f(z) = \sum_{j \in A} a_j z^j,$$

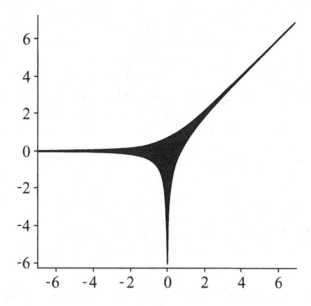

Figure 1.2: The amoeba of $z + w + 1 = 0$.

where $A \subset \mathbb{Z}^n$ is finite, $a_j \in \mathbb{C}$ and $z^j = z_1^{j_1} \ldots z_n^{j_n}$ if $z = (z_1, \ldots, z_n)$ and $j = (j_1, \ldots, j_n)$. Let Δ_f be the *Newton polyhedron* of f, i.e.,

$$\Delta_f = \text{Convex Hull}\{j \in A \mid a_j \neq 0\}.$$

Then

- Every connected component of $\mathbb{R}^n \setminus \mathcal{A}$ is convex.

- The number of components of $\mathbb{R}^n \setminus \mathcal{A}$ is not greater than $\#(\Delta_f \cap \mathbb{Z}^n)$ and not less than the number of vertices of Δ_f.

- There is a naturally defined injection from the set of components of $\mathbb{R}^n \setminus \mathcal{A}$ to $\Delta_f \cap \mathbb{Z}^n$. The vertices of Δ_f are always in the image of this injection. A component of $\mathbb{R}^n \setminus \mathcal{A}$ is bounded if and only if its image is in the interior of Δ_f.

- If $V \subset (\mathbb{C}^*)^2$, then the area of $\mathcal{A}(V)$ is not greater than $\pi^2 \text{Area}(\Delta_f)$.

- If $V \subset (\mathbb{C}^*)^2$ is such that $\text{Area}(\mathcal{A}(V)) = \pi^2 \text{Area}(\Delta_f)$, then the result of coordinatewise multiplication of V by $(c_1, c_2) \in (\mathbb{C}^*)^2$ is the zero set of a polynomial with real coefficients. Furthermore, the real zero set of this polynomial is a real curve with a particular topology in $(\mathbb{R}^*)^2$. If it is non-singular, then its isotopy class in $(\mathbb{R}^*)^2$ depends only on Δ_f. Such curves are called *simple Harnack curves*; historically these were the first examples of curves of degree d in $\mathbb{RP}^2$ with the maximal number of ovals.

- If the coordinates in $(\mathbb{C}^*)^n$ are changed by

$$z_j \mapsto \alpha_j \prod_{k=1}^n z_k^{p_{jk}},$$

with $(p_{jk}) \in \mathrm{GL}(n, \mathbb{Z})$, then the amoeba of V is changed by the affine-linear map of R^n, namely by the linear map corresponding to the matrix (p_{jk}) composed with the translation by $(\log|\alpha_1|, \log|\alpha_2|, \ldots, \log|\alpha_n|)$.

1.2   Families of amoebas

Now, when we are acquainted with amoebas (at least in the case of hypersurfaces), let us consider their families and their limits under an appropriate renormalization.

Let the coefficients of f depend on a parameter t. Let

$$V_t = \{z \in (\mathbb{C}^*)^n \mid \sum_{j \in A \subset \mathbb{Z}^n} a_j(t) \cdot z^j = 0\}$$

be the corresponding variety. To begin let us assume that $a_j(t)$ are polynomial functions in $t > 0$.

Let $A_t(V_t)$ be the amoeba of V_t under the coordinatewise logarithm with base t, i.e.,

$$\mathrm{Log}_t : (\mathbb{C}^*)^n \longrightarrow \mathbb{R}^n,$$
$$(z_1, z_2, \ldots, z_n) \longmapsto (\log_t|z_1|, \log_t|z_2|, \ldots, \log_t|z_n|).$$

We shall see that the limit of $A_t(V_t)$, $t \to \infty$, exists in the Hausdorff metric on compacts of $\mathbb{R}^n$.

Recall the Hausdorff metric

Let A and B be subsets of some metric space (X, d). Then the Hausdorff distance between A and B is given by

$$d_H(A, B) = \max\{d_{\mathrm{asym}}(A, B), d_{\mathrm{asym}}(B, A)\}$$

where d_{asym} denotes the asymmetric Hausdorff distance

$$d_{\mathrm{asym}}(A, B) = \sup_{a \in A} d(a, B),$$

as usual, $d(a, B) = \inf_{b \in B} d(a, b)$. Note that d_H is indeed a metric on the collection of all closed subsets of X. If $d(A, B) = 0$ and both A and B are closed, then $A = B$.

We say that a family $\{A_t\}$, $t \to \infty$, of subsets of X converges in the *Hausdorff metric on compacts* to a set $A \subset X$, if for every compact set $D \subset X$ there exists a neighborhood $U \supset D$ such that $\lim_{t \to \infty} d_H(A_t \cap U, A \cap U) = 0$.

Proposition 1.2. *The family of subsets $\mathcal{A}_t \subset \mathbb{R}^n$ has a limit in the Hausdorff metric on compacts when $t \to \infty$.*

To get an idea of the proof we suggest the following exercise. Let

$$V_t = \{(z, w) \in (\mathbb{C}^*)^2 \mid a(t)z + b(t)w + c(t) = 0\},$$

where $a(t), b(t), c(t)$ are polynomials.

Exercise.

1. If the polynomials a b and c are non-zero constants, then

$$\lim_{t \to \infty} \mathcal{A}_t = Y \subset \mathbb{R}^2,$$

where

$$Y = \{(s, 0) \in \mathbb{R}^2 \mid s \le 0\} \cup \{(0, s) \in \mathbb{R}^2 \mid s \le 0\} \cup \{(s, s) \in \mathbb{R}^2 \mid s \ge 0\}. \quad (1.2)$$

2. If a is a polynomial of degree k, b is a polynomial of degree l and c is a non-zero constant, then

$$\lim_{t \to \infty} \mathcal{A}_t = \tau_{k,l}(Y) \subset \mathbb{R}^2,$$

where

$$\tau_{k,l} : \mathbb{R}^2 \to \mathbb{R}^2, \ (x, y) \mapsto (x - k, y - l)$$

and Y is defined by (1.2).

1.3 Non-Archimedean amoebas

Alternatively we can view a family of algebraic varieties depending on t as a single variety over a field whose elements are functions of t.

The simplest algebraic functions are just polynomials in t. All such polynomials (with complex coefficients) form the ring $\mathbb{C}[t]$. To introduce division we have to pass to a larger ring $\mathbb{C}((t))$ formed by the Laurent power series in t, i.e., functions

$$\phi(t) = \sum_{j=k}^{+\infty} a_j t^j$$

with $a_j \in \mathbb{C}$. Here we may restrict our attention only to the Laurent series $\phi(t)$ that converge in a neighborhood $U \ni 0$. The ring $\mathbb{C}((t))$ is a field, but this field is not algebraically closed. E.g., the equation $z^2 = t$ does not have solutions in $\mathbb{C}((t))$. To make it algebraically closed one has to consider fractional powers of t and the Puiseux series formed by them.

For our purposes it is convenient to allow not only rational but any real powers of t. We define the field $K\{t\}$ of *Puiseux series with real powers locally converging at zero* by

$$K\{t\} = \{\phi : U \to \mathbb{R} \mid \phi(t) = \sum_{j \in I} a_j t^j, \ a_j \in (\mathbb{C}^*), \ t \in U\},$$

where $0 \in U \in \mathbb{R}$ is some open neighborhood of 0 and $I \subset \mathbb{R}$ is a well ordered set (i.e., any subset of I has a minimum). It can be checked that $K\{t\} \supset \mathbb{C}((t))$ is an algebraically closed field.

Furthermore, there exists a non-Archimedean valuation

$$\mathrm{val} : K\{t\} \to \mathbb{R} \cup \{-\infty\},$$

i.e., a map which satisfies the following properties:

1. $\mathrm{val}(f) = -\infty$ if and only if $f = 0$,

2. $\mathrm{val}(fg) = \mathrm{val}(f) + \mathrm{val}(g)$,

3. $\mathrm{val}(f + g) \leq \max\{\mathrm{val}(f), \mathrm{val}(g)\}$.

We define this valuation by

$$\mathrm{val}(\sum_{j \in I} a_j t^j) = -\min(I).$$

As in the degeneration that we considered earlier we have $t \to \infty$, we set $K = K\{\frac{1}{t}\}$ for our function field.

Every valuation defines a norm by

$$|f|_{\mathrm{val}} = e^{\mathrm{val}(f)}.$$

This norm satisfies the stronger, non-Archimedean, version of the triangle inequality

$$|\phi + \psi|_{\mathrm{val}} \leq \max\{|\phi|_{\mathrm{val}}, |\psi|_{\mathrm{val}}\}. \tag{1.3}$$

Remark 1.3. Recall that the Archimedes axiom states that for any numbers $a, b \in \mathbb{R}$ we will have

$$|na| = |a + \cdots + a| > |b|$$

for some number $n \in \mathbb{N}$, where $|a|$ stands for the standard absolute value in $\mathbb{R}$.

If $\phi, \psi \in K$ and $|\phi| < |\psi|$, then

$$|\psi + \cdots + \psi| = |\psi| < |\psi|,$$

so the Archimedes axiom is not satisfied. However, in the modern terminology, being non-Archimedean refers not simply to the absence of the Archimedes axiom, but specifically to the inequality (1.3) that guarantees its absence.

Since $|\cdot|_{\text{val}}$ is the only norm on the Puiseux series in this section, the subscript val to $|\cdot|$ will be omitted from now on.

As in the case of complex varieties, we may use the norm on K to define amoebas of algebraic varieties over K. Let $V_K \subset (K^*)^n$ be an algebraic variety. Again we define the componentwise logarithm of the absolute values

$$\text{Log}_K = \text{Val} : (K^*)^n \to \mathbb{R}^n$$

by

$$\text{Log}_K(\phi_1, \dots, \phi_n) = (\log|\phi_1|, \dots, \log|\phi_n|) = (\text{val}(\phi_1), \dots, \text{val}(\phi_n)).$$

Then $\mathcal{A}(V_K) = \text{Val}(V_K)$ is the (non-Archimedean) amoeba of V_K.

Theorem 1.4. *[a version of Viro patchworking] Let V_t be an algebraic variety with parameter t as defined above and V_K be the corresponding variety in $(K^*)^n$. Then the non-Archimedean amoeba $\mathcal{A}(V_K)$ is the limit of the amoebas $\mathcal{A}(V_t)$ as t goes to infinity with respect to the Hausdorff metric on compacts. In particular, $\lim\limits_{t \to \infty} \mathcal{A}(V_t)$ exists in this sense.*

Theorem 1.5 (Kapranov). *If $V_K \subset (K^*)^n$ is a hypersurface, then the non-Archimedean amoeba $\mathcal{A}(V_K) = \text{Val}(V_K)$ depends only on the valuation of the coefficients of the defining equation for V_K. In other words, if $V_K = \{\sum \alpha_j(t) z^j = 0\}$, $\alpha_j \in K$, then $A(V_K)$ is determined by the values $\text{val}(\alpha_j(t))$.*

As it was noticed by Kapranov [8] the choice of a particular algebraically closed non-Archimedean field K does not affect the geometry of non-Archimedean amoebas as long as the non-Archimedean valuation $K^* \to \mathbb{R}$ is surjective. Another useful choice for such K is provided by the non-standard analysis and considered in the following subsection. Although the following content does not depend on this subsection and uses $K = K\{\frac{1}{t}\}$ as the ground non-Archimedean field, some people might find this other example more intuitive.

1.4   Non-standard complex numbers

Here we present another construction for a non-Archimedean field K with a surjective valuation to $\mathbb{R}$.[1]

We start by recalling of one of the constructions for a generalized limit in analysis. By an *ultrafilter* on the set of natural numbers $\mathbb{N}$, we mean a finitely additive measure v with the following three properties.

- For any set $S \subset \mathbb{N}$, either $v(S) = 0$ or $v(S) = 1$.

- We have $v(S) = 0$, if $S \subset \mathbb{N}$ is finite.

- $v(\mathbb{N}) = 1$.

[1]The author is indebted to M. Kapovich for an illuminating explanation of the asymptotic cone construction in geometry and the relevant point of view on non-standard analysis.

Existence of such υ can be deduced from the axiom of choice.

Definition 1.6. A sequence $a_k \in \mathbb{C}$, $k \in \mathbb{N}$ is called converging to $L \in \mathbb{C}$ with respect to the ultrafilter υ if for any $\epsilon > 0$ we have

$$\upsilon\{k \in \mathbb{N} \mid |a_k - L| \geq \epsilon\} = 0.$$

We say that $a_k \in \mathbb{C}$, $k \in \mathbb{N}$ converges to $L = \infty$ if for any $\epsilon > 0$ we have $\upsilon\{k \in \mathbb{N} \mid |a_k| \leq \epsilon\} = 0$.

We write $\lim\limits_{k \to +\infty}{}^{\upsilon} a_k = L$.

It is easy to see that every sequence of complex number has a limit with respect to υ. E.g., the sequence $0, 1, 0, 1, \ldots$ has the limit 0 or 1 depending on whether the measure of all odd numbers is 1 or the measure of all even numbers is 1 (we should get exactly one of these cases for our υ).

Let $\mathbb{C}^\infty = \{\{a_k\}_{k=1}^{+\infty}\}$ be the set of all sequences $\{a_k\}$ with $a_k \in \mathbb{C}$. Define the equivalence relation by setting $\{a_k\} \sim_\upsilon \{b_k\}$ if $\upsilon\{k \in \mathbb{N} \mid a_k \neq b_k\} = 0$. Let $\mathbb{C}_\upsilon^\infty$ be the set of the corresponding equivalence classes.

We may operate with elements of $\mathbb{C}_\upsilon^\infty$ as with usual complex numbers. We can add them, subtract, multiply and divide coordinatewise. Furthermore, the functions on $\mathbb{C}$ extend to $\mathbb{C}_\upsilon^\infty$ by coordinatewise application. Clearly, $\mathbb{C}_\upsilon^\infty$ is an algebraically closed field.

We say that $\rho = \{\rho_k\} \in \mathbb{C}_\upsilon^\infty$ is positive if $\upsilon\{k \in \mathbb{N} \mid \rho_k > 0\} = 1$ (as usual, $\rho_k > 0$ implies, in particular, that $\rho_k \in \mathbb{R}$). We say that ρ is infinitely large if $\lim\limits_{k \to +\infty}{}^{\upsilon} \rho_k = \infty$. Let us fix once and for all a positive infinitely large ρ.

The inequality $|a| < \rho^N$ for $a = \{a_k\} \in \mathbb{C}_\upsilon^\infty$ means that $\upsilon\{k \in \mathbb{N} \mid |a_k| \leq \rho_k^N\} = 1$. Define

$$A = \{a \in \mathbb{C}_\upsilon^\infty \mid \exists N \in \mathbb{N} : |a| < \rho^N\}.$$

Similarly, we define

$$B = \{a \in \mathbb{C}_\upsilon^\infty \mid |a| < \rho^N \; \forall N \in \mathbb{Z}\}.$$

Define $\mathbb{C}_\upsilon^\rho = A/B$. Clearly, it is an algebraically closed field. Furthermore,

$$\mathrm{Log}_\rho : \mathbb{C}_\upsilon^\rho \to \mathbb{R} \cup \{-\infty\}, \; a \mapsto \log_\rho(|a|) = \{\log_{\rho_k} |a_k|\}$$

is a surjective valuation.

The field $\mathbb{C}_\upsilon^\rho$ can be considered as a field of non-standard complex numbers and we have just seen that it is a non-Archimedean field. The choice of $K = \mathbb{C}_\upsilon^\rho$ allows us to consider the map $\mathrm{Val} = \mathrm{Log}_\rho$. This map is a limiting map in the family Log_t and the limit can be obtained by substitution of the infinitely large value $t = \rho$.

1.5 The tropical semifield $\mathbb{T}$

Definition 1.7. We set the *tropical semifield* $\mathbb{T} = \mathbb{R} \cup \{-\infty\}$ to be the real numbers enhanced with $-\infty$ and equipped with the following arithmetic operations (we use quotation marks to distinguish them from the classical arithmetic operations with real numbers)

$$\text{``}x + y\text{''} = \max\{x, y\},$$
$$\text{``}xy\text{''} = x + y.$$

It is easy to check that $\mathbb{T}$ is a commutative semigroup with respect to addition (here $-\infty \in \mathbb{T}$ plays the rôle of the additive zero), a commutative group with respect to multiplication (here $0 \in \mathbb{T}$ plays the rôle of the multiplicative unit) and that we have the distribution law

$$\text{``}x(y + z) = xy + xz\text{''}.$$

In other words, $\mathbb{T}$ is a true semifield. This semifield drastically lacks subtraction: as tropical addition is an idempotent operation, we have "$x + x = x$".

Remark 1.8. The term *tropical* is borrowed from Computer Science (where it was introduced to commemorate a Brazilian scientist Imre Simon). These arithmetic operations under different names appeared even earlier. E.g., Litvinov and Maslov [36] used the term *idempotent analysis* and related the process of passing from the classical arithmetics to the tropical arithmetics to the *quantization* process of Schrödinger, but in the opposite direction. This is the source for another name for passing to the tropical limit, "*dequantization*".

We finish this remark with the corresponding deformation of arithmetic operations on $\mathbb{T}$ from "classical" to "tropical" (cf. e.g., [36], [72]). Let $\mathbb{R}_{\geq 0}$ be the semifield of non-negative real numbers equipped with classical arithmetic operations. The map

$$\log_t : \mathbb{R}_{\geq 0} \to \mathbb{T},$$

$t > 1$, induces some arithmetic operations on $\mathbb{T}$. Namely, we have

$$x \oplus_t y = \log_t(t^x + t^y), \quad x \otimes_t y = x + y$$

for $x, y \in \mathbb{T}$. Clearly, for any finite $t > 0$ the set $\mathbb{T}$ equipped with these operations is a semifield isomorphic to $\mathbb{R}_{\geq 0}$ (the isomorphism is provided by $\log_t$ itself). In particular, the semifields $(\mathbb{T}, \oplus_t, \otimes_t)$ are mutually isomorphic. However, we have

$$\lim_{t \to +\infty} x \oplus y = \text{``}x + y\text{''}, \quad x \otimes_t y = \text{``}xy\text{''},$$

thus the tropical semifield $\mathbb{T}$ (which not isomorphic to $\mathbb{R}_{\geq 0}$ as $\mathbb{T}$ is idempotent and $\mathbb{R}_{\geq 0}$ is not) is the limit of a family of semifields isomorphic to $\mathbb{R}_{\geq 0}$.

Figure 1.3: The graph of a tropical parabola $a_0 + a_1x + a_2x^2$ and the graph of the corresponding function $j \mapsto a_j$.

Tropical polynomials and corresponding tropical hypersurfaces

We do not have subtraction in $\mathbb{T}$, but we do not need it to define polynomials, as a polynomial is a sum of monomials. Denote $\mathbb{Z}_{\geq 0} = \mathbb{N} \cup \{0\}$.

Definition 1.9. Let $A \subset (\mathbb{Z}_{\geq 0})^n$ be finite and $a_j \in \mathbb{T}$ for all $j \in A$. Then a tropical polynomial is given by

$$f(x) = \text{``}\sum_{j \in A} a_j x^j\text{''} = \max_{j \in A}(a_j + \langle j, x \rangle), \tag{1.4}$$

$x \in \mathbb{T}^n$.

Remark 1.10. Equation (1.4) may recall the definition of the *Legendre transform*. Recall that for an arbitrary function $\varphi : \mathbb{R}^n \to \mathbb{R}$ its Legendre transform L_φ is defined by

$$L_\varphi(x) = \max_j(\langle j, x \rangle - \varphi(j)).$$

This definition makes sense even if φ is defined only on a subset $A \subset \mathbb{R}^n$ — we just take the maximum over $j \in A$ or, equivalently, extend φ to $\mathbb{R}^n$ by setting $\varphi(j) = +\infty$, $j \notin A$. The function L_φ is convex even if φ is not convex. However, if φ is not convex, then there exists a (non-strictly) convex function

$$\tilde{\varphi} : \mathbb{R}^n \to \mathbb{R} \cup \{+\infty\}$$

such that $L_{\tilde{\varphi}} = L_\varphi$. It is easy to see that the graph of such $\tilde{\varphi}$ can be obtained from the overgraph of ϕ by taking the convex hull.

We have the tropical polynomial f equal to the Legendre transform of the function $j \mapsto -a_j$ defined on the finite subset $A \subset \mathbb{R}^n$. Since A is finite the function f is a piecewise-linear convex function.

Our next step is to define hypersurfaces associated to tropical polynomials. The neutral element with respect to addition is $-\infty$, but tropical polymomials almost never take value $-\infty$. Because of the lack of subtraction we have to be very careful in phrasing the classical definition of hypersurface in order to make this definition work also for $\mathbb{T}$. For such definition one can take the definition of the zero locus of a polynomial as the locus where the multiplicative inverse of the polynomial is not regular.

Definition 1.11. Let f be a tropical polynomial. Then the corresponding tropical hypersurface V_f is given by

$$V_f = \{x \in \mathbb{R}^n \mid f \text{ is not smooth in } x\} \subset \mathbb{R}^n = (\mathbb{T}^*)^n.$$

It is also called the corner locus of f.

Indeed, near V_f we have "$\frac{0}{f}$" not locally convex and, therefore, not regular.

Example 1.12. The locus of the tropical parabola from Figure 1.3 is formed by the projections onto the x-axis of the two corners of the graph. One can compute that these points are $a_0 - a_1$ and $a_1 - a_2$ if $2a_1 > a_0 + a_1$. Otherwise this locus consists of one *double* point $\frac{a_0 - a_2}{2} \in \mathbb{T}$.

Remark 1.13. For every polynomial f in $K[x_1, \dots, x_n]$ we can define its *tropicalization* by replacing the Puiseux series coefficients from K with their valuations from $\mathbb{T}$.

Kapranov's Theorem 1.5 may be now reformulated in the following way. *The hypersurface of the tropicalization of a polynomial over K coincides with its (non-Archimedean) amoeba.*

While the study of tropical hypersurfaces (and, as a matter of fact, also complete intersections) is relatively easy, even the definition of general tropical varieties takes time (see [45] and [46]). In the next section we give some basic treatment of general tropical varieties in dimension 1, i.e., *tropical curves*. We treat them as abstract tropical varieties, i.e., in a manner analogous to the Riemann surfaces). The geometric structure serving as a tropical counterpart of complex structure turns out to be integer affine structure.

1.6 Tropical curves and integer affine structure

We start with a preliminary definition that is well-known in classical geometry.

Definition 1.14. Let M be a manifold. An integer affine structure on M is given by an open covering $\{U_j\}$ of M with embedding charts

$$\varphi_j : U_j \longrightarrow \mathbb{R}^n,$$

such that every overlapping transition map

$$\varphi_k \circ \varphi_j^{-1} : \varphi_j(U_j \cap U_k) \to \varphi_k(U_j \cap U_k)$$

can be extended to a map $\mathbb{R}^n \to \mathbb{R}^n$ obtained by the composition of a map

$$\Phi_{kj} : \mathbb{R}^n \to \mathbb{R}^n,$$

linear over $\mathbb{Z}$, and a translation by an arbitrary vector in $\mathbb{R}^n$, in other words, by an element of $\mathrm{GL}_n(\mathbb{Z}) \ltimes \mathbb{R}^n$.

A manifold M equipped with an integer affine structure is called a $\mathbb{Z}$-*affine manifold*. Let M and N be two $\mathbb{Z}$-affine manifolds (of dimensions $\dim M$ and $\dim N$), and let $f : M \to N$ be a map. We say that f is a $\mathbb{Z}$-*affine map* if for every $x \in M$ there exist charts $U_j \ni x$ and $U_k^{(N)} \ni f(x)$ such that $\phi_k^{(N)} \circ f \circ \phi^{-1}$ can be extended to a map $\mathbb{R}^{\dim M} \to \mathbb{R}^{\dim N}$ obtained by the composition of a map $\mathbb{R}^{\dim M} \to \mathbb{R}^{\dim N}$, linear over $\mathbb{Z}$, and a translation by an arbitrary vector in $\mathbb{R}^{\dim N}$.

Given an integer affine structure on M, one can make it a full affine structure by including in the set of charts all $\mathbb{Z}$-affine embeddings $M \supset U \to \mathbb{R}^n$. Recall that an (integer) affine structure is called *complete* if for every chart $\varphi_j : U_j \to \mathbb{R}^n$ and $y \in \mathbb{R}^n$ there exists a finite sequence of charts $U_j = U_{j_0}, \ldots, U_{j_l}$ such that $U_{j_{m-1}} \cap U_{j_m} \neq \emptyset$ and $\phi_{j_l}(U_l) \ni y$.

Note that since $U_{j_{m-1}} \cap U_{j_m} \neq \emptyset$, affine-linear maps $\Phi_{j_m j_{m-1}} : \mathbb{R}^n \to \mathbb{R}^n$ are defined. One can check that the composition

$$\tilde{\Phi}_j(y) = \phi_{j_l}^{-1} \circ \Phi_{j_m j_{m-1}} \circ \ldots \Phi_{j_1 j_0}(y) \in M$$

does not depend on the choice of the chart sequence and gives a well-defined covering map $\tilde{\Phi}_j : \mathbb{R}^n \to M$ called the *developing map*, cf. [66].

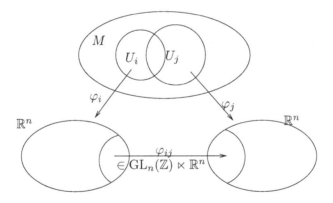

Figure 1.4: Illustrating the $\mathbb{Z}$-affine structure.

We have a well-defined notion of an *integer tangent vector* to a $\mathbb{Z}$-affine manifold M. These are the vectors corresponding to vectors in $\mathbb{Z}^n$ under the differential of charts ϕ_j. We say that an integer tangent vector is *primitive* if it is not a nontrivial (not ± 1) integer multiple of another integer tangent vector.

Note that specifying a $\mathbb{Z}$-affine structure on a 1-manifold is equivalent to specifying a metric. Indeed, we have

$$\mathrm{GL}_n(\mathbb{Z}) = O(1);$$

the metric is specified by setting the primitive tangent vector (which is unique up to the sign) to have a unit length.

The simplest class of examples of $\mathbb{Z}$-affine manifolds is provided by quotients $\mathbb{R}^n/\Lambda$ where Λ is a *lattice* in $\mathbb{R}^n$, i.e., a discrete subgroup of $\mathbb{R}^n$ isomorphic to $\mathbb{Z}^n$. These examples play the role of tropical tori; some of them admit a polarization and thus are tropical Abelian varieties (see [47]).

However, most tropical varieties of dimension n are not topologically manifolds. They are polyhedral complexes of dimension n which still come equipped with an integer affine structure. In this section we look only at the case $n = 1$. Topologically tropical curves are graphs. Compact tropical curves are finite graphs.

Let Γ be a finite graph. Instead of defining a $\mathbb{Z}$-affine structure in the non-manifold case for a general polyhedral complex (see [45]), we take advantage of dimension 1 where we can express a $\mathbb{Z}$-affine structure by using a metric. Denote

$$\Gamma^\circ = \Gamma\backslash\{1 - \text{valent vertices}\}. \tag{1.5}$$

Definition 1.15. A compact tropical curve is a connected finite graph Γ equipped with a complete inner metric on Γ°.

Thus the length of an edge of Γ is infinite if and only if this edge is adjacent to a 1-valent vertex. The integer affine structure near a k-valent vertex $x \in \Gamma$ with $k > 1$ can be thought as an isometric map

$$\phi : U \to Y \subset \mathbb{R}^{k-1} \tag{1.6}$$

of a neighborhood $U \ni x$ to the subspace $Y \subset \mathbb{R}^{k-1}$ obtained as the union of the negative part of the $k - 1$ axes of $\mathbb{R}^{k-1}$ (considered with the metric induced by the Euclidean metric on $\mathbb{R}^n$) and the ray $R = \{(t,\ldots,t) \in \mathbb{R}^{k-1} \mid t \geq 0\}$, where the metric is induced by the Euclidean metric on $\mathbb{R}^n$ scaled by $\sqrt{k-1}$ (so that the length of the primitive integer vector $(1,\ldots,1)$ is unity).

The restriction of tropical polynomials from $\mathbb{R}^{k-1}$ to Y define regular functions on open sets of Γ. Together they form *the structure sheaf* $\mathcal{O}$.

Note that if Γ is a 3-valent graph with n univalent vertices, then the number of finite edges of Γ is equal to $3(\dim H_1(\Gamma))-3+n$ if $2g+n > 2$, where $g = \dim H_1(\Gamma)$. Furthermore, all tropical curves with n *marked* (i.e., numbered) 1-valent vertices and with $\dim H_1(\Gamma) = g$ form *the tropical moduli space* $\mathcal{M}_{g,n}^{\text{trop}}$ that can be naturally compactified. It can be shown that this compactification is a tropical orbifold (of dimension $3g - 3 + n$) as long as $2g + n > 2$. Furthermore, if $g = 0$, then it is a manifold.

Definition 1.16. The number $g = b_1(\Gamma)$ is called *the genus* of a tropical curve Γ.

As in the complex case the genus g can be interpreted as the dimension of regular 1-forms on Γ (see [47]). All regular forms can be used to form the *Jacobian variety* of Γ. As in classical geometry one has the tropical counterpart of the Abel–Jacobi theorem. The Riemann–Roch theorem holds in the form of an inequality and can be used to give a lower bound for the dimension of the space of deformations of a tropical curve.

Many other (but not all, see e.g., [54]) classical theorems for complex curves also have their tropical counterpart. As we will not need them in the applications of tropical geometry considered in the next two chapters we just refer the reader to [45], [46], [47]. To relate abstract tropical curves to these applications we finish this section by looking at tropical maps $h : \Gamma^\circ \to \mathbb{R}^n$, where Γ° is defined by (1.5).

Such a map h is tropical if it is given by a $\mathbb{Z}$-affine map in every small chart (1.6). The following proposition translates this to the language of metric graphs.

Proposition 1.17. *Let Γ be a compact tropical curve and Γ° be its finite part. A map*

$$h : \Gamma^\circ \to \mathbb{R}^n$$

is tropical if and only if the following two conditions hold.

- *For every edge $E \subset \Gamma^\circ$, $h|_E : E \to \mathbb{R}^n$ is a smooth map such that dh maps every unit tangent vector to E to an integer vector in $\mathbb{R}^n$.*

- *For every k-valent vertex $x \in \Gamma^\circ$ denote with $v_1, \ldots, v_k$ the outgoing unit tangent vectors to the edges $E_1, \ldots, E_k$ adjacent to x. Then we have*

$$\sum_{j=1}^{k} (dh|_{E_j})_x v_j = 0.$$

Clearly for every edge $E \subset \Gamma^\circ$ the image $W \subset \mathbb{Z}^n$ of the unit tangent vector is the same for all points of E (up to multiplication by -1). The largest natural divisor $w \in \mathbb{N}$ of W (i.e., the GCD of its coordinates) is called *the weight* of the image $h(E)$. In the remaining part of our discussion we will be looking at the images $h(\Gamma^\circ) \subset \mathbb{R}^2$ and using them for the needs of classical real and complex geometry.

1.7 Exercises

Exercise 1.1. Let $V \subset (\mathbb{C}^*)^2$ be given by $z + w + 1 = 0$.

- Write down explicit inequalities defining the amoeba $\mathrm{Log}(V) \subset \mathbb{R}^2$.

- Write down explicit inequalities defining the image $\mu(V) \subset \mathbb{R}^2$, where $\mu : (\mathbb{C}^*)^2 \to \mathbb{R}^2$ is the moment map given by

$$(z, w) \mapsto \left(\frac{|z|^2}{1 + |z|^2 + |w|^2}, \frac{|w|^2}{1 + |z|^2 + |w|^2} \right).$$

- Prove that $\mathrm{Log}\big|_{(\mathbb{R}_+)^2}$ and $\mu\big|_{(\mathbb{R}_+)^2}$ are both diffeomorphisms onto their images, and find these images.

- Prove that, for any convex lattice polytope $\Delta \subset \mathbb{R}^n$, the moment map

$$\mu_\Delta(x) = \frac{\sum_{\omega \in \Delta \cap \mathbb{Z}^n} x^\omega \cdot \omega}{\sum_{\omega \in \Delta \cap \mathbb{Z}^n} x^\omega}$$

is a real analytic diffeomorphism of the positive orthant $\mathbb{R}^n_+$ onto the interior $\mathrm{Int}(\Delta)$ of Δ.

Exercise 1.2. Find a family $\{A_t\}$, $t \to +\infty$ of subsets of $\mathbb{R}$, a set $A \subset \mathbb{R}$, and a set $U \subset \mathbb{R}$ such that

$$\lim_{t \to +\infty} d_H(A_t, A) = 0 \quad \text{and} \quad \lim_{t \to +\infty} d_H(A_t \cap U, A \cap U) = +\infty.$$

Exercise 1.3. Denote by $\mathbb{K}$ the field of formal power series $\sum_{r \in J} a_r t^r$ with complex coefficients, where $J \subset \mathbb{R}$ is well-ordered. Let $V \subset (\mathbb{K}^*)^2$ be the non-Archimedean curve defined by the equation $1 + z + w + tzw = 0$, and let $V_f \subset \mathbb{R}^2$ be the tropical curve given by the tropical polynomial $f(x,y) = $ "$0 + x + y + 1 \cdot x \cdot y$". Prove that $\mathrm{Log}_\mathbb{K}(V) = V_f$.

Exercise 1.4. Let K be a field with a non-Archimedean valuation $\mathrm{val} : K^* \to \mathbb{R}$, $A \subset \mathbb{Z}^2$ a non-empty finite set, $F(z,w) = \sum_{(i,j) \in A} a_{ij} z^i w^j$ a Laurent polynomial over K. Show that the set

$$\mathcal{A}(F) = \{(\mathrm{val}(z), \mathrm{val}(w)) \ : \ F(z,w) = 0, \ z, w \in K^*\}$$

is contained in the tropical curve V_f defined by the tropical polynomial $f(x) = \max_{j \in A}(a_j + \langle j, x \rangle)$. Assuming in addition that K is algebraically closed of characteristic zero, and the valuation val is dominant (surjective), prove that $\mathcal{A}(F)$ is dense in (coincides with) V_f.

Exercise 1.5. (A research problem.) Can you find a tropical curve in $\mathbb{R}^3$ that is the limit of amoebas of real rational algebraic curves which are knotted (*e.g.*, realize a trefoil knot)?

Chapter 2

Patchworking of algebraic varieties

2.1 Introduction: A general idea of the patchworking construction

Consider the diagram

$$X \lhook\joinrel\longrightarrow Y \tag{2.1}$$
$$(\mathbb{C}, 0)$$

where Y is a germ of a 1-parameter flat[1] family of algebraic varieties with $\dim Y \geq 3$, such that the fibres Y_t are reduced irreducible for $t \neq 0$, and the central fibre Y_0 splits into a few reduced components. Further on, X is a germ of a 1-parameter flat family of subvarieties $X_t \subset Y_t$ for $t \in (\mathbb{C}, 0)$. When considering the diagram (2.1) over the reals, we assume that X and Y are equipped with a complex conjugation which commutes with the projections, and we restrict the parameter range to $t \in [0, \tau)$, $\tau > 0$, taking the respective preimages in X and Y.

In this situation one can state the following two problems (below K denotes either the complex or the real field):

---

[1] The flatness over a smooth one-dimensional base means that the projection is a proper analytic surjection.

Degeneration (tropicalization) problem

Data: Flat families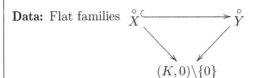

Aim: Find appropriate limit fibres $X_0 \subset Y_0$ defined over the origin.

Deformation (patchworking) problem

Data: A flat family $Y \longrightarrow (K, 0)$ with reduced irreducible fibres Y_t for $t \neq 0$ and reduced reducible central fibre $Y_0 = \bigcup_{k=1}^{N} Y_{0,k}$, and a subvariety $X_0 \subset Y_0$.

$$
\begin{array}{ccc}
X_0 \lhook\joinrel\longrightarrow Y_0 \lhook\joinrel\longrightarrow Y \\
\downarrow \qquad\quad \downarrow \\
0 \lhook\joinrel\longrightarrow (K, 0)
\end{array}
$$

Aim: Restore a flat family $X \longrightarrow (K, 0)$ so that the fibres X_t will inherit certain properties of X_0 for $t \neq 0$.

As the result of the patchworking construction, the subvariety X_t of Y_t for $t \neq 0$ appears to be "glued" out of the components $X_{0,k} = X_0 \cap Y_{0,k}$ of X_0, $k = 1, 2, \ldots, N$.

The aim of this chapter is to demonstrate deep links of the patchworking method with tropical geometry, in particular, to show how the patchworking applies to tropical enumerative geometry.

We have tried to present the material in a logically connected way, with basic ideas and constructions, illustrating examples, exercises. Some proofs or details are skipped, since they can be found in the recommended literature:

- for Section "2.2. Elements of toric geometry" we refer to [12];

- for Section 2.3. "Viro's patchworking method" we refer to [67, 68, 70, 71, 72] as well as to [16, 19, 20, 21, 25, 26];

- for Section 2.4. "Patchworking singular algebraic hypersurfaces" we refer to [56];

- for Section 2.5. "Tropicalization and patchworking in the enumeration of nodal curves" we refer to [42, 58, 59].

For other applications of the patchworking construction we refer to [24, 55, 57, 63].

2.2 Elements of toric geometry

Definition 2.1. A complex n-dimensional *toric variety* is an irreducible (usually normal) algebraic variety X with an action of the torus $(\mathbb{C}^*)^n$ having an open dense orbit (isomorphic just to $(\mathbb{C}^*)^n$).

Toric varieties include, in particular, the affine and projective spaces, the rational geometrically ruled surfaces and some other important varieties.

2.2.1 Construction of toric varieties

A rational convex polyhedral cone is a set

$$
\sigma = \left\{ \sum_{i \in I} \tau_i \cdot v_i \in \mathbb{R}^n \mid \tau_i \geq 0, i \in I \right\} ,
$$

where I is a finite index set, and $v_i \in \mathbb{Z}^n$ for $i \in I$ are given vectors, the so-called generators. The cone σ is called strongly convex if it does not contain a linear subspace of positive dimension.

The dual cone

$$
\sigma^\vee = \left\{ \phi \in (\mathbb{R}^n)^* \mid \phi(u) \geq 0, u \in \sigma \right\}
$$

is a rational convex polyhedral cone as well.

For every strongly convex rational cone, the set $S_\sigma = \sigma^\vee \cap \mathbb{Z}^n$ equipped with the componentwise addition is a finitely generated semigroup (Gordan's lemma). Therefore the set

$$
\mathbb{C}[S_\sigma] \stackrel{\text{def}}{=} \left\{ \sum_{i \in \tilde{S}} a_i z^i \ \middle| \ \tilde{S} \subset S_\sigma \text{ finite and } a_i \in \mathbb{C}, z = (z_1, z_2, \ldots, z_n) \right\}
$$

is a finitely-generated polynomial algebra. Hence the variety $\mathcal{U}_\sigma = \operatorname{Spec} \mathbb{C}[S_\sigma]$ associated to the cone σ is an affine algebraic variety with $\dim \mathcal{U}_\sigma = \dim \sigma^\vee$. It is always irreducible and normal. Furthermore, it can be described as a subvariety of an affine space in the following way: choose semigroup generators $v_1, v_2, \ldots, v_s \in S_\sigma$, then $U_\sigma \subset \mathbb{C}^s$ is defined by the equations

$$
y_1^{n_1} y_2^{n_2} \cdots y_s^{n_s} = 1
$$

for any relation

$$
n_1 \cdot v_1 + n_2 \cdot v_2 + \cdots + n_s \cdot v_s = 0
$$

with integers $n_1, n_2, \ldots, n_s$.

Example 2.2. Let $\sigma = \{0\} \subset \mathbb{R}^n$. Then $\sigma^\nu = \mathbb{R}^n$ and the semigroup $S_\sigma = \mathbb{Z}^n$ is generated by the finite set

$$\{e_1, e_2, \ldots, e_n, -e_1, -e_2, \ldots, -e_n\}.$$

Thus, we obtain the polynomial algebra $\mathbb{C}[S_\sigma] = \mathbb{C}[x_1, x_1^{-1}, x_2, x_2^{-1}, \ldots, x_n, x_n^{-1}]$ and the affine toric variety $\mathcal{U}_\sigma \simeq (\mathbb{C}^*)^n = \operatorname{Spec} \mathbb{C}[S_\sigma]$, which can be defined by the equations

$$y_1 y_2 = 1, \ y_3 y_4 = 1, \ \ldots, \ y_{2n-1} y_{2n} = 1$$

in $\mathbb{C}^{2n}$. The action of the torus $T = (\mathbb{C}^*)^n$ is given by an action on the polynomial algebra $\mathbb{C}[S_\sigma]$, which is defined by

$$(\mathbb{C}^*)^n \times \mathbb{C}[S_\sigma] \longrightarrow \mathbb{C}[S_\sigma],$$

$$(\lambda_1, \lambda_2, \ldots, \lambda_n) \cdot (x_1, x_2, \ldots, x_n) \longmapsto (\lambda_1^{-1} x_1, \lambda_2^{-1} x_2, \ldots, \lambda_n^{-1} x_n).$$

Example 2.3. Let $\tau \subset \sigma$ be a face of a cone σ, then

$$
\begin{array}{ccc}
\tau \subset \sigma & & \mathcal{U}_\tau \hookrightarrow \mathcal{U}_\sigma \\
\Big\Downarrow & & \Big\Uparrow \\
\tau^\nu \supset \sigma^\nu & \Longrightarrow & \mathbb{C}[S_\tau] \hookleftarrow \mathbb{C}[S_\sigma]
\end{array}
$$

where the embeddings in the right-hand side are compatible with the $(\mathbb{C}^*)^n$-action, and $\mathcal{U}_\tau$ is an open subset in $\mathcal{U}_\sigma$. In particular, for a strongly convex cone σ, the set $\{0\} \subset \sigma$ is a face and hence $(\mathbb{C}^*)^n$ is an open subset of $\mathcal{U}_\sigma$. Moreover, in fact, it is dense.

2.2.2   A toric variety associated with a fan

Definition 2.4. A fan is a finite collection $\mathcal{F}$ of strongly convex rational polyhedral cones, such that

- for any $\sigma \in \mathcal{F}$, each face τ of σ belongs to F, and

- for any two cones $\sigma_1, \sigma_2 \in \mathcal{F}$, the intersection $\sigma_1 \cap \sigma_2$ is a common face.

Definition 2.5. Let $\mathcal{F}$ be a fan. Then the associated toric variety $\operatorname{Tor}(\mathcal{F})$ is defined as the quotient

$$\operatorname{Tor}(\mathcal{F}) = \bigcup_{\sigma \in \mathcal{F}} \mathcal{U}_\sigma \Big/ \sim \, ,$$

that is the union of the affine toric varieties, associated with the faces σ as in Example 2.3, modulo an equivalence relation $\sim$. The relation identifies for every pair of incident faces $\tau \subset \sigma \in \mathcal{F}$ the variety $\mathcal{U}_\tau$ with its embedding into $\mathcal{U}_\sigma$.

2.2.3 A toric variety associated with a convex lattice polyhedron

Let $\Delta \subset \mathbb{R}^n$ be a convex lattice polyhedron, we construct a cone for any face $\tau \subset \Delta$. The cone of τ,

$$C_\tau(\Delta) = \langle u - u' \mid u \text{ a vertex of } \Delta, u' \text{ a vertex of } \tau \rangle,$$

is generated by the integral difference vectors of vertices from τ and Δ. Then the fan corresponding to Δ is

$$\mathcal{F}(\Delta) = \{C_\tau(\Delta)^\nu \mid \tau \subset \Delta\}.$$

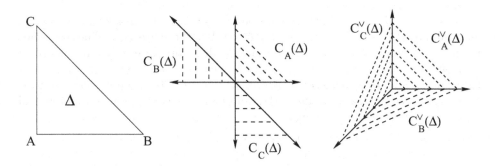

Figure 2.1: Fan of a polytope and its associated cones.

The corresponding variety $\mathrm{Tor}(\Delta) = \mathrm{Tor}(\mathcal{F}(\Delta))$ is called the toric variety associated with the polyhedron Δ. Its structure reflects the combinatorics of the polyhedron Δ and can be described as follows.

Theorem 2.6. *There is a 1-to-1 correspondence between the orbits of the $(\mathbb{C}^*)^n$-action on $\mathrm{Tor}(\Delta)$ and the faces of Δ including Δ itself, such that:*

1. $\dim \mathcal{O}_\sigma = \dim \sigma$, *where $\mathcal{O}_\sigma$ is the orbit corresponding to the face σ,*

2. $\overline{\mathcal{O}}_\sigma \simeq \mathrm{Tor}(\sigma)$,

3. $\overline{\mathcal{O}}_\sigma \supset \mathcal{O}_\tau \Longleftrightarrow \tau$ *is a face of σ.*

Here $\overline{\mathcal{O}}_\sigma$ means the closure of $\mathcal{O}_\sigma$.

Remark 2.7. The group of affine-linear automorphisms $\mathrm{Aff}(\mathbb{Z}^n)$ of $\mathbb{Z}^n$ acts on the set of convex lattice polyhedra and thus induces isomorphisms of toric varieties.

Example 2.8. Let $\Delta \subset \mathbb{R}^2$ be the triangle with vertices $(0,0), (1,0)$ and $(0,1)$. Then $\mathrm{Tor}(\Delta) \simeq \mathbb{C}P^2$, and the sides of Δ correspond to the coordinate lines in $\mathbb{C}P^2$ (further on, the divisors on a toric surface $\mathrm{Tor}(\Delta)$, corresponding to the sides of Δ, will be called **boundary divisors**), whereas the vertices of Δ correspond to the fundamental points in $\mathbb{C}P^2$ (see Figure 2.1).

2.2.4   Embedding of $\mathrm{Tor}(\Delta)$ into a projective space

Define a map φ on $(\mathbb{C}^*)^n \subset \mathrm{Tor}(\Delta)$ by

$$\varphi : (\mathbb{C}^*)^n \longrightarrow \mathbb{C}P^N, \quad \varphi(z) = (z^{\omega_0} : z^{\omega_1} : \cdots : z^{\omega_N}),$$

where[2]

$$N = |\Delta \cap \mathbb{Z}^n| - 1, \quad \Delta \cap \mathbb{Z}^n = \{\omega_1, \omega_2, \ldots, \omega_N\}.$$

It extends up to a regular map of $\mathrm{Tor}(\Delta)$:

$$\varphi : \mathrm{Tor}(\Delta) \xrightarrow{\ \sim\ } \overline{\varphi\left((\mathbb{C}^*)^n\right)} \hookrightarrow \mathbb{C}P^N.$$

In case $n = 2$, the map φ is always embedding, i.e., the linear system on $\mathrm{Tor}(\Delta)$, generated by the monomials z^i, $i \in \Delta \cap \mathbb{Z}^n$, (called the **tautological linear system**) is very ample. In case $n \geq 3$, it is so, provided that, for any face $\sigma \subset \Delta$ (including Δ itself), the points $\sigma \cap \mathbb{Z}^n$ generate the lattice $L(\sigma) \cap \mathbb{Z}^n$, where $L(\sigma) \subset \mathbb{R}^n$ is the minimal affine subspace containing σ. Replacing Δ by the polytope $m\Delta$ with a sufficiently large m, we can make the tautological linear system very ample and thus can assume that φ is an embedding.

Example 2.9. Let τ_d^n be an n-dimensional simplex with side length d, i.e.,

$$\tau_d^n = \mathrm{conv}\{0, d \cdot e_1, d \cdot e_2, \ldots, d \cdot e_n\} \subset \mathbb{R}^n,$$

where e_i denotes the i-th unit vector in $\mathbb{R}^n$. Then $\mathrm{Tor}(\tau_d^n) \simeq \mathbb{C}P^n$ and

$$\varphi : \mathrm{Tor}(\tau_d^n) = \mathbb{C}P^n \longrightarrow \mathbb{C}P^N$$

is the d-multiple Segre embedding.

2.2.5   The real part of a toric variety and the moment map

The same construction over $\mathbb{R}$ leads to the notion of a real toric variety $\mathrm{Tor}_{\mathbb{R}}(\Delta) \subset \mathrm{Tor}(\Delta)$. It contains as an open dense subset the real torus

$$(\mathbb{R}^*)^n = \coprod_{\varepsilon \in \{\pm 1\}^n} (\mathbb{R}^*)_\varepsilon^n,$$

which can be written as the disjoint union of the open orthants $(\mathbb{R}^*)_\varepsilon^n \subset \mathbb{R}^n$. We then define

$$\mathrm{Tor}_{\mathbb{R},\varepsilon}(\Delta) \overset{\mathrm{def}}{=} \overline{(\mathbb{R}^*)_\varepsilon^n} \subset \mathrm{Tor}_{\mathbb{R}}(\Delta).$$

The following nice geometric description of $\mathrm{Tor}_{\mathbb{R}}(\Delta)$ is based on the moment map.

---

[2]Here and further on, the notation |finite set| means the cardinality of the given finite set.

Definition 2.10. The moment map $\mu = \mu_\Delta : (\mathbb{C}^*)^n \to \Delta$ is defined by

$$\mu(z) = \frac{\sum_{i \in \Delta \cap \mathbb{Z}^n} i \cdot |z|^i}{\sum_{i \in \Delta \cap \mathbb{Z}^n} |z|^i} \qquad \text{for } z \in (\mathbb{C}^*)^n \subset \text{Tor}(\Delta), \tag{2.2}$$

where $|z|^i = |z_1|^{i_1} \cdot |z_2|^{i_2} \cdots \cdots |z_n|^{i_n}$. It continuously extends up to

$$\mu_\Delta : \text{Tor}(\Delta) \longrightarrow \Delta.$$

The following statement is well known.

Theorem 2.11. *For any $\varepsilon \in \{\pm 1\}^n$, the restricted moment map*

$$\mu_\Delta : \text{Tor}_{\mathbb{R},\varepsilon}(\Delta) \longrightarrow \Delta$$

is a homeomorphism, such that

1. *$\mu_\Delta : (\mathbb{R}^*)^n_\varepsilon \longrightarrow \text{Int}(\Delta)$ is an analytic diffeomorphism and*

2. *for any face $\sigma \subset \Delta$ the set*

$$\text{Tor}_{\mathbb{R},\varepsilon}(\sigma) \overset{def}{=} \text{Tor}(\sigma) \cap \text{Tor}_{\mathbb{R},\varepsilon}(\Delta)$$

is mapped by μ_Δ homeomorphically onto $\sigma \subset \Delta$.

Thus, the real toric variety $\text{Tor}_{\mathbb{R}}(\Delta)$ is glued up from 2^n copies $\text{Tor}_{\mathbb{R},\varepsilon}(\Delta)$ of Δ by the following identification. Two faces $\text{Tor}_{\mathbb{R},\varepsilon}(\sigma)$ and $\text{Tor}_{\mathbb{R},\delta}(\sigma)$ for $\varepsilon, \delta \in \{\pm 1\}^n$ are identified, if either, for any $i \in \sigma \cap \mathbb{Z}^n$ it holds that $\varepsilon^i = \delta^i$, or, for any $i \in \sigma \cap \mathbb{Z}^n$ it holds that $\varepsilon^i = -\delta^i$.

Remark 2.12. In formula (2.2), we can replace the right-hand side by an expression

$$\mu(z) = \frac{\sum_{i \in F} i \cdot \lambda_i |z|^i}{\sum_{i \in F} \lambda_i |z|^i} \qquad \text{for } z \in (\mathbb{C}^*)^n \subset \text{Tor}(\Delta),$$

where $F \subset \Delta$ is finite including all the vertices of Δ, the numbers λ_i are positive, and the statement of Theorem 2.11 will remain valid.

Assume that Δ lies entirely in the closed non-negative orthant $\mathbb{R}^n_+$, and, for any $\varepsilon \in \{\pm 1\}^n$, denote by Δ_ε the copy of Δ in the closed orthant $\mathbb{R}^n_\varepsilon$ obtained by the reflection $r_\varepsilon : \mathbb{R}^n_+ \to \mathbb{R}^n_\varepsilon$ with respect to suitable coordinate hyperplanes. Then we define the following modification of the moment map:

$$\mu_{\Delta,\varepsilon} \overset{def}{=} r_\varepsilon \circ \mu_\Delta : \text{Tor}_{\mathbb{R},\varepsilon}(\Delta) \to \Delta_\varepsilon .$$

Introduce also the map

$$\psi_\Delta : \bigcup_{\varepsilon \in \{\pm 1\}^n} \Delta_\varepsilon \to \text{Tor}_{\mathbb{R}}(\Delta) ,$$

which is inverse to $\mu_{\Delta,\varepsilon}$ on each part $\mathrm{Tor}_{\mathbb{R},\varepsilon}(\Delta)$, $\varepsilon \in \{\pm 1\}^n$, and is onto. Geometrically, ψ_Δ glues together the 2^n symmetric copies Δ_ε of Δ along their faces by the rule described in Remark 2.12.

We shall also use the complexified moment map, defined as follows. Let $\Delta \subset \mathbb{R}^n$ be an n-dimensional convex lattice polytope. Decompose the complex torus $(\mathbb{C}^*)^n$ as $(\mathbb{R}^*)^n_+ \times (S^1)^n$ by

$$(z_1, \ldots, z_n) \quad \leftrightarrow \quad \left((|z_1|, \ldots, |z_n|), \left(\frac{z_1}{|z_1|}, \ldots, \frac{z_n}{|z_n|} \right) \right) ,$$

and define the complexified polytope $\mathbb{C}\Delta \stackrel{\text{def}}{=} \Delta \times (S^1)^n$. The complexified moment map $\mathbb{C}\mu_\Delta : (\mathbb{C}^*)^n \to \mathbb{C}\Delta$ is given by

$$\mathbb{C}\mu_\Delta(z_1, \ldots, z_n) = \left(\mu_\Delta(z_1, \ldots, z_n), \left(\frac{z_1}{|z_1|}, \ldots, \frac{z_n}{|z_n|} \right) \right) \in \mathbb{C}\Delta .$$

Theorem 2.11 immediately yields

Theorem 2.13. *The map $\mathbb{C}\mu_\Delta$ is a real analytic diffeomorphism of $(\mathbb{C}^*)^n$ onto $\mathrm{Int}(\Delta) \times (S^1)^n \subset \mathbb{C}\Delta$. Furthermore, the inverse map naturally extends up to a surjection $\mathbb{C}\psi : \mathbb{C}\Delta \to \mathrm{Tor}(\Delta)$ such that, for each proper face $\sigma \subset \Delta$,*

1. *$\mathbb{C}\psi$ takes $\sigma \times (S^1)^n \subset \mathbb{C}\Delta$ onto $\mathrm{Tor}(\sigma) \subset \mathrm{Tor}(\Delta)$;*

2. *the restriction $\mathbb{C}\psi : \mathrm{Int}(\sigma) \times (S^1)^n \to (\mathbb{C}^*)^{\dim \sigma} \subset \mathrm{Tor}(\sigma)$ induces a fibration with fibre $(S^1)^{\mathrm{codim}\,\sigma}$.*

2.2.6 Hypersurfaces in toric varieties

A hypersurface in $\mathrm{Tor}(\Delta)$, which does not contain the boundary divisors, is uniquely determined by its intersection with the torus $(\mathbb{C}^*)^n$, and that intersection can be defined by an equation,

$$f(z) := \sum_{i \in \Delta \cap \mathbb{Z}^n} a_i z^i = 0 , \tag{2.3}$$

containing at least two monomials. We restore the original hypersurface by taking the closure $\overline{\{f = 0\}} \cap (\mathbb{C}^*)^n \subset \mathrm{Tor}(\Delta)$ in the toric variety $\mathrm{Tor}(\Delta)$. We can, if necessary, replace Δ by $N \cdot \Delta$ with $N \in \mathbb{N}$, since $\mathrm{Tor}(\Delta) \simeq \mathrm{Tor}(N \cdot \Delta)$.

Consider an algebraic hypersurface defined in $(\mathbb{C}^*)^n$ by equation (2.3). The closure $\overline{\{f = 0\}} \cap (\mathbb{C}^*)^n \subset \mathrm{Tor}(\Delta)$ is an algebraic hypersurface in the toric variety $\mathrm{Tor}(\Delta)$. The intersection of this hypersurface with the subvarieties $\mathrm{Tor}(\sigma)$, σ being a proper face of Δ, can be described in the following way:

$$\overline{\{f = 0\}} \cap \mathrm{Tor}(\sigma) = \{f^\sigma = 0\} ,$$

where $f^\sigma = \sum_{i \in \sigma \cap \mathbb{Z}^n} a_i \cdot z^i$ is the *truncation* of f to σ. More generally, let $\Delta \subset \Delta$ be the Newton polytope of f. To recover the intersection $\overline{\{f = 0\}} \cap \mathrm{Tor}(\sigma)$, we cannot just take the restriction of f to a face σ, since maybe no integer point of this face corresponds to a monomial of f, i.e., $\sigma \cap \Delta' = \emptyset$. Instead, assuming for simplicity that $\sigma \subset \Delta$ is a facet (face of codimension 1), we take the exterior normal vector $v \in \mathbb{Z}^n$ of σ and then choose the face $\sigma' \subset \Delta'$, where the functional

$$(\mathbb{R}^n \ni u \longmapsto u \cdot v)|_{\Delta'}$$

attains its maximum. Then we get

$$\{f = 0\} \cap \mathrm{Tor}(\sigma) = \{f^{\sigma'} = 0\}.$$

2.3 Viro's patchworking method

In 1979–80, Viro suggested a patchworking construction for obtaining real non-singular projective algebraic hypersurfaces with prescribed topology. This method was a major breakthrough in Hilbert's 16th problem, and since then it has been the most efficient constructive method in this field. Viro immediately solved the isotopy classification problem for plane real curves of degree 7 and made significant progress in degree 8 (up to degree 5 it has been known since the 19th century, the degree 6 case was completed by Gudkov in 1969). So far almost all known topological types of real nonsingular algebraic curves and, more generally, real nonsingular projective hypersurfaces have been constructed using Viro's patchworking method or its modifications.

2.3.1 Chart of a real polynomial

Let $\Delta \subset \mathbb{R}^n_+$ be an n-dimensional convex lattice polytope.

Definition 2.14. The *chart* $\mathrm{Chart}_{\Delta,\varepsilon}(f)$ of a polynomial $f \in \mathbb{R}[z_1, \ldots, z_n]$ in the polytope Δ_ε is given by

$$\mathrm{Chart}_{\Delta,\varepsilon}(f) = \overline{\mu_{\Delta,\varepsilon}(\{f = 0\} \cap (\mathbb{R}^*)^n_\varepsilon)} \subset \Delta_\varepsilon.$$

The **complex chart** $\mathbb{C}\,\mathrm{Chart}_\Delta(f)$ of a polynomial $f \in \mathbb{C}[z_1, \ldots, z_n]$ in the complexified polytope $\mathbb{C}\Delta$ is defined as

$$\mathbb{C}\,\mathrm{Chart}_{\Delta,\varepsilon}(f) = \overline{\mathbb{C}\mu_\Delta(\{f = 0\} \cap (\mathbb{C}^*)^n)} \subset \mathbb{C}\Delta\,.$$

Example 2.15. Let $f(x, y) = xy - x - y + 1$. The Newton polygon Δ of f is the square with the vertices $(0,0)$, $(0,1)$, $(1,0)$, and $(1,1)$. Then the affine curve $\{f = 0\} \subset \mathbb{R}^2$ and the charts of f in the polygons $\Delta_{++} = \Delta$, Δ_{+-}, Δ_{-+}, and Δ_{--} look as shown in Figure 2.2.

The following properties of charts will be important for us.

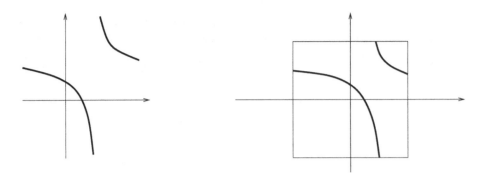

Figure 2.2: Chart of a real polynomial.

Lemma 2.16. *Let Δ be the Newton polytope of a polynomial f. Then for any face $\sigma \subset \Delta$,*

$$\mathrm{Chart}_{\Delta,\varepsilon}(f) \cap \sigma_\varepsilon = \mathrm{Chart}_{\sigma,\varepsilon}(f^\sigma), \quad \varepsilon \in \{\pm 1\}^n \ ,$$

where $f^\sigma(z) = \sum_{i \in \sigma \cap \mathbb{Z}^n} a_i z^i$. More generally, if

- *$\Delta' \subset \Delta$ is the Newton polytope of f,*
- *$\sigma \subset \Delta$ is a facet of Δ,*
- *π is an exterior normal vector of σ,*
- *$\varphi(u) = \langle \pi, u \rangle$ is a linear functional, and*
- *$\sigma' \subset \Delta'$ is the face of Δ', where φ attains its maximum,*

then

$$\mathrm{Chart}_{\Delta,\varepsilon}(f) \cap \sigma_\varepsilon = \mathrm{Chart}_{\sigma,\varepsilon}(f^{\sigma'}), \quad \varepsilon \in \{\pm 1\}^n \ .$$

A similar statement is valid for the complex charts. To avoid unnecessary complications, we formulate it in a simplified form (which however can be extended up to a general case in view of Remark 2.7):

Lemma 2.17. *Let $\Delta \subset \mathbb{R}^n$ be an n-dimensional Newton polytope of a polynomial f, and let σ be a k-dimensional face of Δ, lying in a coordinate k-plane. Then*

$$\mathbb{C}\,\mathrm{Chart}_\Delta(f) \cap (\sigma \times (S^1)^n)) \simeq \mathbb{C}\,\mathrm{Chart}_\sigma(f^\sigma) \times (S^1)^{n-k} \ .$$

Lemma 2.18. *Assume that $f \in \mathbb{R}[z_1, \dots, z_n]$ is **completely non-degenerate**, that is neither f nor its truncation f^σ to any face σ of its Newton polytope Δ have singular points in the torus $(\mathbb{C}^*)^n$. Then the chart $\mathrm{Chart}_{\Delta,\varepsilon}(f)$ is a hypersurface in Δ_ε with boundary on $\partial\Delta_\varepsilon$, which is smooth in the interior of Δ_ε and, for $n \geq 3$, has corners on the boundary.*

If in addition, $\dim \Delta = n$, then $\mathbb{C}\,\mathrm{Chart}_\Delta(f)$ is a codimension 2 submanifold in $\mathbb{C}\Delta$ with boundary on $\partial\mathbb{C}\Delta = \partial\Delta \times (S^1)^n$. The chart $\mathbb{C}\,\mathrm{Chart}_\Delta(f)$ is smooth in $\mathrm{Int}(\mathbb{C}\Delta)$ and, for $n \geq 3$, has corners on the boundary.

2.3.2　Patchworking of real nonsingular hypersurfaces

We present here three versions of the Viro patchworking theorem, real affine, real projective, and equivariant ones, and give the proof of the real affine version, leaving to the reader the proof of the remaining versions.

　　Let us be given the following data:

- an n-dimensional convex lattice polytope $\Delta \subset \mathbb{R}^n_+$;

- a convex piecewise-linear function $\nu : \Delta \to \mathbb{R}$, whose linearity domains form a subdivision

$$\Delta = \Delta_1 \cup \Delta_2 \cup \cdots \cup \Delta_N$$

　 into convex lattice polytopes (such a subdivision is called **convex**);

- a set of numbers $a_i \in \mathbb{R}$ for $i \in \Delta \cap \mathbb{Z}^n$, such that $a_i \neq 0$ for any vertex i of the above subdivision, and the polynomials

$$f_m(z) = \sum_{i \in \Delta_m \cap \mathbb{Z}^n} a_i z^i, \quad m = 1, \ldots, N,$$

　 are completely non-degenerate.

Theorem 2.19 (Real affine patchworking). *Under the above assumptions, for any* $\varepsilon \in \{\pm 1\}^n$ *and any sufficiently small* $t > 0$, *the chart* $\mathrm{Chart}_{\Delta, \varepsilon}(f_{(t)})$ *of the Viro polynomial*

$$f_{(t)} = \sum_{i \in \Delta} a_i t^{\nu(i)} \cdot z^i$$

is isotopic in Δ_ε *to*

$$\bigcup_{m=1}^{N} \mathrm{Chart}_{\Delta_m, \varepsilon}(f_m) \tag{2.4}$$

by an isotopy, which leaves the faces of Δ_ε *invariant.*

Theorem 2.20 (Real projective patchworking). *Under the above assumptions, for any sufficiently small* $t > 0$, *the (singular in general) manifolds*

$$\psi_\Delta \left(\bigcup_{\varepsilon \in \{\pm 1\}^n} \mathrm{Chart}_{\Delta, \varepsilon}(f_{(t)}) \right) \subset \mathrm{Tor}_{\mathbb{R}}(\Delta)$$

and

$$\bigcup_{m=1}^{N} \psi_\Delta \left(\bigcup_{\varepsilon \in \{\pm 1\}^n} \mathrm{Chart}_{\Delta, \varepsilon}(f_m) \right) \subset \mathrm{Tor}_{\mathbb{R}}(\Delta) \tag{2.5}$$

are isotopic in $\mathrm{Tor}_{\mathbb{R}}(\Delta)$ *by an isotopy, which leaves the real toric divisors* $\mathrm{Tor}_{\mathbb{R}}(\sigma) \subset \mathrm{Tor}_{\mathbb{R}}(\Delta)$ *invariant for all faces* $\sigma \subset \Delta$.

Theorem 2.21 (Equivariant patchworking). *Under the above assumption, for any sufficiently small $t > 0$, the (singular in general) manifolds*

$$\mathbb{C}\psi_\Delta \left(\mathbb{C}\operatorname{Chart}_\Delta(f_{(t)}) \right)$$

and

$$\mathbb{C}\psi_\Delta \left(\bigcup_{m=1}^N \mathbb{C}\operatorname{Chart}_{\Delta_m}(f_m) \right) \tag{2.6}$$

are equivariantly isotopic in $\operatorname{Tor}(\Delta)$ by an isotopy which leaves the toric subvarieties $\operatorname{Tor}(\sigma)$, $\sigma \subset \Delta$, invariant.

Remark 2.22. From Lemmas 2.16, 2.17, and 2.18 we can deduce that the charts $\operatorname{Chart}_{\Delta_m,\varepsilon}(f_m)$ and $\operatorname{Chart}_{\Delta_l,\varepsilon}(f_l)$, as well as $\mathbb{C}\operatorname{Chart}_{\Delta_m}(f_m)$ and $\mathbb{C}\operatorname{Chart}_{\Delta_l}(f_l)$ agree along the common face $\sigma_\varepsilon = \Delta_{m,\varepsilon} \cap \Delta_{l,\varepsilon}$, respectively, along $\sigma \times (S^1)^n = \mathbb{C}\Delta_m \cap \mathbb{C}\Delta_l$. Furthermore, all the charts $\operatorname{Chart}_{\Delta_m,\varepsilon}(f_m)$, $m = 1, 2, \ldots, N$, glue up into a manifold with corners and boundary on $\partial\Delta_\varepsilon$, and the same holds for the charts $\mathbb{C}\operatorname{Chart}_{\Delta_m}(f_m)$, $m = 1, \ldots, N$.

Proof of Viro's theorem. First, we slightly deform the function ν in the space of convex piecewise-linear functions determining the same subdivision of Δ in order to get ν defined over $\mathbb{Q}$. This is always possible, since the latter space is an open rational cone. Notice also that such a deformation of ν does not affect the isotopy type of the real and complex charts of the Viro polynomials $f_{(t)}$ as $t > 0$ is small enough. Indeed, for a fixed ν, the charts of $f_{t)}$ are isotopic in some interval $0 < t < t^*$, and the upper bound t^* (corresponding to a singular hypersurface $\{f_{(t^*)} = 0\} \cap (\mathbb{C}^*)^n$) depends analytically on the coordinates of the vertices of the graph of ν.

Afterwards we multiply ν by a large integer in order to provide $\nu(\Delta \cap \mathbb{Z}^n) \subset \mathbb{Z}$ (such a multiplication simply means the change of parameter in (2.4)).

Now consider the convex polytope

$$\widetilde{\Delta} = \{(x, y) \in \mathbb{R}^{n+1} \ : \ x \in \Delta, \nu(x) \leq y \leq M\},$$

where $M > \max\nu|_\Delta$ (see Figure 2.3). Then $f_{(t)}$ and $t - c$ with $c \in \mathbb{R}$ define hypersurfaces in $\operatorname{Tor}(\widetilde{\Delta})$.

First, the hypersurface $\{t - c = 0\}$ is for $c \neq 0$ isomorphic to $\operatorname{Tor}(\Delta)$, where Δ can be interpreted as the upper facet of $\widetilde{\Delta}$. The isomorphism is provided by a family of straight lines $z = \bar{c}$ for $\bar{c} \in (\mathbb{C}^*)^n$, respectively compactified, which transversally cross $\operatorname{Tor}(\Delta)$ each at one point, and also cross $\{t - c = 0\}$ transversally each at one point. Notice that $\{t - c = 0\}$ crosses $\operatorname{Tor}(\sigma) \subset \operatorname{Tor}(\widetilde{\Delta})$ only for lateral faces $\sigma \in \widetilde{\Delta}$. When $c \to 0$, the hypersurface $\{t - c = 0\}$ degenerates into

$$\bigcup_{m=1}^N \operatorname{Tor}(\widetilde{\Delta}_m),$$

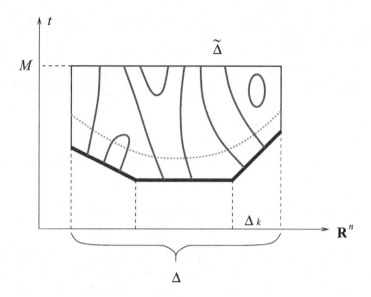

Figure 2.3: Proof of Viro's theorem.

where $\widetilde{\Delta}_1, \widetilde{\Delta}_2, \ldots, \widetilde{\Delta}_N$ are the faces of the graph of ν (shown as the thick broken line designating the bottom of $\overline{\Delta}$ in Figure 2.3), that is $\mathrm{Chart}_{\widetilde{\Delta},\varepsilon}(t-c)$, $c > 0$, (shown as dotted graph in Figure 2.3) converges to $\bigcup_{m=1}^{N} \widetilde{\Delta}_{m,\varepsilon}$ as c tends to zero.

On the other hand the hypersurface $\{f_{(t)} = 0\}$ (whose chart consists of the remaining eight graphs above the thick lines in Figure 2.3) crosses $\{t - c = 0\}$ transversally as $c > 0$ sufficiently small and, furthermore, $\{f_{(t)} = 0\}$ crosses $\bigcup_{m=1}^{N} \mathrm{Tor}(\widetilde{\Delta}_m)$ along $\bigcup_{m=1}^{N}\{f_{(t)}^{\widetilde{\Delta}_m} = 0\}$. Notice that, due to $\nu(\mathbb{Z}^n) \subset \mathbb{Z}$, the projection of $\widetilde{\Delta}_m$ onto Δ_m extends up to an affine automorphism of $\mathbb{Z}^{n+1}$. Thus it defines an isomorphism

$$\mathrm{Tor}(\widetilde{\Delta}_m) \simeq \mathrm{Tor}(\Delta_m),$$

which takes $\{f_{(t)}^{\widetilde{\Delta}_m}\}$ to $\{f_m = 0\}$. The latter can be viewed as

$$f_{(t)}^{\widetilde{\Delta}_m} = \sum_{i \in \Delta_m} a_i t^{\alpha_0 + \alpha \cdot i} z^i = t^{\alpha_0} \cdot f_m(z_1 \cdot t^{\alpha_1}, z_2 \cdot t^{\alpha_2}, \ldots, z_n \cdot t^{\alpha_n})$$

where

$$\nu\big|_{\Delta_m}(i) = \alpha_0 + \alpha \cdot i = \alpha_0 + \alpha_1 i_1 + \cdots + \alpha_n i_n \ .$$

One can show that $\{f_{(t)} = 0\}$ crosses $\bigcup_{m=1}^{N} \mathrm{Tor}(\widetilde{\Delta}_m)$ transversally. Hence the following diagram holds.

$$
\begin{array}{ccc}
\mathrm{Chart}_{\widetilde{\Delta},\varepsilon}(f_{(t)}) \cap \mathrm{Chart}_{\widetilde{\Delta},\varepsilon}(t-c) & \xrightarrow{\;c \to 0\;} & \displaystyle\bigcup_{m=1}^{N} \mathrm{Chart}_{\widetilde{\Delta}_m,\varepsilon}(f_{(t)}^{\widetilde{\Delta}_m}) \\[2mm]
\{t-c=0\}\simeq\mathrm{Tor}(\Delta) \Big\updownarrow & & \Big\updownarrow \text{projection} \\[2mm]
\mathrm{Chart}_{\Delta,\varepsilon}(f_{(c)}) \subset \Delta_\varepsilon & & \displaystyle\bigcup_{m=1}^{N} \mathrm{Chart}_{\Delta_m,\varepsilon}(f_m)
\end{array}
$$

Now we are done, since $f_1, f_2, \ldots, f_N, f_{(t)}$ are completely non-degenerate. $\square$

Example 2.23. We illustrate the Viro patchworking theorem by applying it to the classification of nonsingular real plane curves of degree 6. By the Harnack inequality (see, for instance, [69]), a real nonsingular plane projective curve of degree m may have at most $(m-1)(m-2)/2+1$ connected components. The real non-singular curves of a given degree having that maximal number of connected components are called M-curves. For an even m all the connected components of a nonsingular real curve are ovals, that is null-homologous circles in $\mathbb{R}P^2$. So, a real plane nonsingular sextic curve may have at most 11 ovals, and, due to the Gudkov–Rokhlin congruence (again we refer to the survey [69]), the M-sextics must have one of the following isotopy types $\langle 9 + 1\langle 1 \rangle \rangle$, $\langle 1 + 1\langle 9 \rangle \rangle$, or $\langle 5 + 1\langle 5 \rangle \rangle$, i.e., they have one oval embracing 1, 5, or 9 ovals, and outside respectively 9, 5, or 1 ovals (see Figure 2.4). Using the Viro patchworking construction we shall show that there exist sextic curves of each of the three above isotopy types.

We divide the Newton polygon $\Delta = \mathrm{conv}\{(0,0), (6,0), (0,6)\}$ of a generic plane curve of degree 6 into two triangles $\Delta_1 = \mathrm{conv}\{(0,0), (0,3), (6,0)\}$ and $\Delta_2 = \mathrm{conv}\{(0,3), (0,6), (6,0)\}$ (clearly, this subdivision lifts up to the graph of a convex piecewise-linear function on Δ). Then we find coefficients a_{kj}, $k, j \geq 0$, $k + j \leq 6$, so that the corresponding polynomials f_1 and f_2 have charts as shown in Figure 2.4. In fact, it is enough to build a polynomial f_1 realizing one of the required charts, and then the other polynomials can be obtained from that one by suitable coordinate changes (see Figure 2.5).

2.3.3   Combinatorial patchworking

The combinatorial patchworking is a rather simple, but very powerful, particular case of the Viro method.

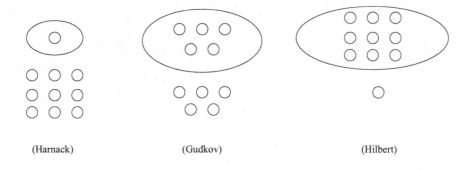

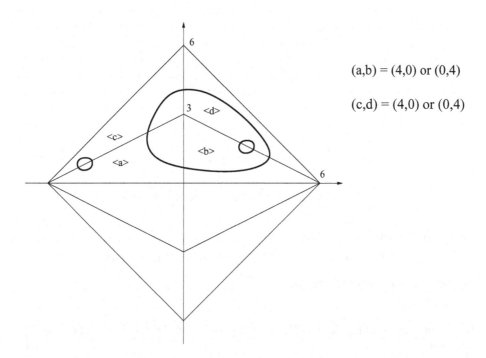

$$(a,b) = (4,0) \text{ or } (0,4)$$

$$(c,d) = (4,0) \text{ or } (0,4)$$

Figure 2.4: The M-curves of degree 6.

Let $\Delta \subset \mathbb{R}^n$ be an n-dimensional lattice simplex and

$$f(z) = \sum_{i \in \Delta \cap \mathbb{Z}^n} a_i z^i$$

a real polynomial such that $a_i \neq 0$ if and only if i is a vertex of Δ. Then f is called

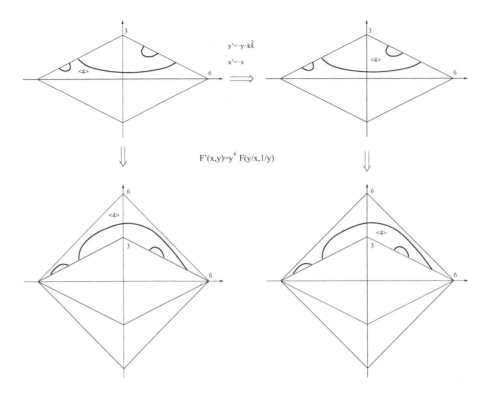

Figure 2.5: Construction of the block polynomials.

(a non-degenerate) $(n + 1)$ - *nomial.* Let

$$T_1^n = \text{conv}\{(0, \ldots, 0), (1, 0, \ldots, 0), \ldots, (0, \ldots, 0, 1)\}$$

be the *unit simplex* of dimension n and

$$\Phi : \Delta \longrightarrow T_1^n$$

be an affine map taking Δ onto T_1^n. It induces a rational map of $\text{Tor}(\Delta)$ to $\text{Tor}(T_1^n)$, which in turn reduces to a diffeomorphism $\Phi_* : \text{Tor}_{\mathbb{R},+}(\Delta) \to \text{Tor}_{\mathbb{R},+}(T_1^n)$ taking $\text{Chart}_{\Delta,+}(f) \subset \Delta$ onto $\text{Chart}_{T_1^n,+}(H) \subset T_1^n$, where H is a linear polynomial, defining a real hyperplane in $\mathbb{P}^n$. The topology of the chart of a hyperplane and its disposition with respect to the faces of T_1^n depend only on the signs of the coefficients: namely, it is isotopic to a hyperplane section in T_1^n, which separates the vertices of different signs (see Figure 2.6).

For any other chart $\text{Chart}_{\Delta,\varepsilon}(f)$, $\varepsilon = (\varepsilon_1, \ldots, \varepsilon_n) \in \{\pm 1\}^n$, we make the coordinate change $(z_1, \ldots, z_n) \mapsto (\varepsilon_1 z_1, \ldots, \varepsilon_n z_n)$, sending the orthant $(\mathbb{R}^*)^n_\varepsilon$ to $(\mathbb{R}^*)^n_+$ and redefining the signs of the coefficients according to the rule

$$\text{sign}\, a_{i,\varepsilon} = \text{sign}\, a_i \cdot \varepsilon^i, \quad i \in \Delta \cap \mathbb{Z}^n \,, \tag{2.7}$$

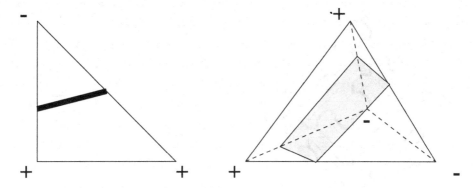

Figure 2.6: Charts of an $(n + 1)$-nomial.

and construct the corresponding chart as above.

The combinatorial patchworking is the following procedure. Assume that, in the initial data of the patchworking construction, $\Delta = \Delta_1 \cup \cdots \cup \Delta_N$ is a triangulation, and the numbers $a_i \in \{-1, 0, 1\}$ are non-zero if and only if i is a vertex $\Delta_1, \ldots, \Delta_N$. Then by Theorems 2.19 and 2.20, the polynomial $f_{(t)}$ realizes the gluing of charts of $(n + 1)$-nomials. Namely, we take the subdivision $\Delta = \Delta_1 \cup \cdots \cup \Delta_N$, and put at each vertex the corresponding coefficient a_i. Then reflecting with respect to the coordinate hyperplanes, we extend the subdivision to each polytope Δ_ε, $\varepsilon \in \{\pm 1\}^n$, as well as the distribution of signs at the vertices with the rule (2.7). Finally, in every simplex $\Delta_{k,\varepsilon}$ we take the hyperplane section, which hits the midpoint of each edge, which joins vertices with different signs. The resulting piecewise-linear complex is isotopic to a chart of the Viro polynomial $f_{(t)}$. The real algebraic hypersurfaces obtained in a combinatorial patchworking are called T-**hypersurfaces** (T-**curves** in the planar case).

We illustrate the use of the combinatorial patchworking in two examples.

Example 2.24. In 1876, Harnack constructed, using small deformations of reducible curves, a series of plane projective M-curves of any degree. In fact, all these M-curves can be obtained in the combinatorial patchworking. Take the Newton triangle $T_d^2 = \text{conv}\{(0,0), (d,0), (0,d)\}$ of a general plane curve of degree d, and consider its primitive convex triangulation, i.e., a subdivision into lattice triangles of Euclidean area $1/2$ (in the n-dimensional case, a primitive triangulation is that into lattice simplices of the minimal Euclidean volume, equal to $1/n!$). Define the coefficients as follows:

$$a_{kj} = \begin{cases} 1, & \text{if } k \equiv j \equiv 0 \mod 2, \\ -1, & \text{otherwise} \end{cases}, \quad (k, j) \in T_d^2 \cap \mathbb{Z}^2 .$$

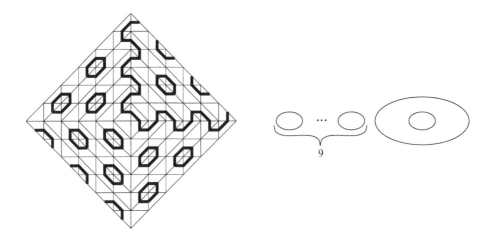

Figure 2.7: The Harnack M-sextic of type $\langle 9 + 1\langle 1 \rangle \rangle$.

The result of the patchworking is a Harnack M-curve (see Figure 2.7), and its isotopy type does not depend on the choice of a primitive convex triangulation of T_d^2.

Example 2.25. As another application of the combinatorial patchworking we present the construction of a real plane curve of degree 10 found by I. Itenberg in 1993 (see Figure 2.8), which contradicts the Ragsdale conjecture formulated in 1906.

2.3.4 A tropical point of view on the combinatorial Viro patchworking

Let $A \subset \mathbb{Z}^n$ be a finite set and

$$y = p(x) = \sum_{i \in A} a_i x^i \quad , \quad x \in \mathbb{R}_+^n$$

a real polynomial with positive coefficients. To compare it with a tropical polynomial, we pass to logarithmic coordinates, i.e.,

$$v = \log y, \quad u = \log x.$$

In this coordinate system the equation $y = p(x)$ turns into

$$v = L_p(u) = \log \left(\sum_{i \in A} e^{iu + b_i} \right)$$

with $a_i = e^{b_i}$. Now we introduce the tropical polynomial

$$M_p(u) = \max_{i \in A} \{iu + b_i\}.$$

Figure 2.8: A counterexample to Ragsdale's conjecture: curve of degree 10 with real scheme $\langle 29 \rangle + 1\langle 2 + 1\langle 2 \rangle \rangle$.

It holds that (see Figure 2.9)

$$M_p(u) \leq L_p(u) = \log \left(\sum_{i \in A} e^{iu + b_i} \right) \leq \log \left(|A| \cdot \max_{i \in A} e^{iu + b_i} \right) = M_p(u) + \log |A|.$$

To get a better approximation, we consider a family of polynomials, namely, the Viro polynomials

$$p_{(t)} = \sum_{i \in A} a_i t^{\nu(i)} \cdot x^i \quad \text{with } t > 0.$$

Passing to logarithmic coordinates, introducing a new parameter $h = -\frac{1}{\log t} > 0$ and applying the homothety

$$u_1 = u \cdot h \qquad v_1 = v \cdot h \ ,$$

we similarly obtain

$$p_{(t)}(u_1) = h \cdot \log \left(\sum_{i \in A} e^{\frac{iu_1 - \nu(i) + b_i h}{h}} \right)$$

with $t^{\nu(i)} = e^{\log(t) \cdot \nu(i)} = e^{\frac{-\nu(i)}{h}}$. This polynomial converges point-wise to

$$M_{P_{(t)}}(u_1) = \max_{i \in A}(i \cdot u_1 - \nu(i))$$

as $h \to 0$, or, equivalently, as $t \to 0$.

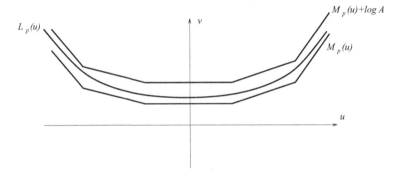

Figure 2.9: Tropicalization of a real polynomial.

Furthermore, let us be given the following data:

- a convex piecewise-linear function $\nu : \mathrm{conv}(A) \to \mathbb{R}$, and
- a general real polynomial

$$p_{(t)}(x) = \sum_{\substack{i \in A \\ a_i > 0}} a_i t^{\nu(i)} x^i - \sum_{\substack{i \in A \\ a_i < 0}} (-a_i) t^{\nu(i)} x^i.$$

Letting

$$p_{(t),+} = \sum_{a_i > 0} a_i t^{\nu(i)} x^i, \quad p_{(t),-} = \sum_{a_i < 0} (-a_i) t^{\nu(i)} x^i \ ,$$

and using the preceding convergence result, we immediately obtain that the real algebraic hypersurface $\{p_{(t)} = 0\}$ in the positive orthant $\mathbb{R}^n_+$, after the coordinate change described above, is well approximated (in fact, converges to as $t \to 0$) by the projection of the intersection of the graphs of the tropical polynomials $M_{p(t),+}$ and $M_{p(t),-}$ as t tends to zero. This, in fact, is another point of view on the combinatorial patchworking of real algebraic hypersurfaces. Recall that a close representation of the combinatorial patchworking, based on the use of amoebas and their limits, was proposed in Theorem 1.4, Chapter 1.

2.3.5 Patchworking of pseudo-holomorphic curves on ruled surfaces

Consider the patchworking data, given in Section 2.3.2. The convexity of the subdivision $\Delta = \Delta_1 \cup \cdots \cup \Delta_N$, needed in the proof of the patchworking Theorems 2.19, 2.20, and 2.21, is a restriction, i.e., there are non-convex subdivisions (see an example in Figure 2.10 borrowed from [6]).

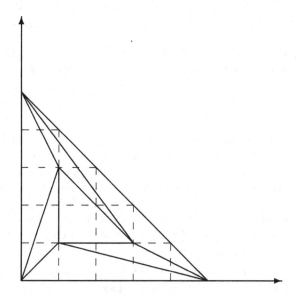

Figure 2.10: A non-convex triangulation.

On the other hand, the unions of charts (2.4), (2.5), and (2.6) are piece-wise smooth manifolds with boundary for any subdivision of Δ. It is natural to ask what structure can be defined on such manifolds if the subdivision is not convex. We intend to show that, if $\text{Tor}(\Delta)$ is the projective plane or a geometrically ruled rational surface, then the union of charts as above is (equivariantly in the real case) isotopic to a pseudo-holomorphic curve.

We, first, recall some necessary definitions and facts about pseudo-holomorphic curves. A Riemann surface M embedded in X (where X stands for $\mathbb{C}P^2$ or a rational ruled surface Σ_m, $m \geq 0$) is called a **pseudo-holomorphic curve**, if it is a J-holomorphic curve in some tame almost complex structure J on X (see [17]).

If in addition $\mathrm{Conj}_* \circ J = J^{-1} \circ \mathrm{Conj}_*$ and $\mathrm{Conj}(M) = M$, where $\mathrm{Conj} : X \to X$ is the standard real structure on X (this is the real structure of X as a real toric surface associated with a polygon), then we call M a **real pseudo-holomorphic curve**.

Pseudo-holomorphic curves and pencils of lines. To verify that the constructed surfaces are pseudo-holomorphic curves, we use the criterion suggested in [10, 49]. Blowing up the plane at one of the fundamental points of the coordinate system, we replace it by Σ_1. Thus, let X be Σ_m, $m \geq 0$, $E \subset X$ be such that $E^2 = -m$, and let $\pi : X \to E$ be the respective ruling. According to [49], Section 5.3, and [10], Section 4.1, an oriented smooth surface $M \subset X$, homologous to $aE + bF$ (F being a fibre of π, where $a > 0$, $b \geq ma$) is a pseudo-holomorphic curve if any fibre $F_c = \pi^{-1}(c)$, $c \in E$, except for a finite number of them, crosses M transversally at a points with multiplicity $+1$, and any of the remaining fibres F_c crosses M at $a - 2$ points with multiplicity $+1$ and at one more point p_c so that, up to a fibre-wise homotopy in a neighborhood of p_c in X, one has in suitable local coordinates $F_c = \{z = 0\}$ and $M = \{z = w^2\}$.

In what follows we construct a fibration of X over $\mathbb{C}P^1$ with spherical fibres, all but finitely many of which cross V, the union of charts (2.6), transversally as required above; after a small isotopy, the fibration becomes smooth and symplectomorphic to π, and the remaining intersections of the fibres with V become as required.

Refinement of subdivision. Let $X = \Sigma_m$, $m \geq 0$, and let $a > 0$, $b \geq \max\{ma, 1\}$ be integers. Denote by Δ' the trapeze (or triangle) with vertices $(0, 1)$, $(a, 1)$, $(0, b + 1)$, and $(a, b - ma + 1)$ (see Figure 2.11(a)), and by Δ the trapeze with vertices $(0, 0)$, $(a, 0)$, $(0, b + 1)$, and $(a, b - ma + 1)$ (see Figure 2.11(b)). Clearly, $\mathrm{Tor}(\Delta) = \Sigma_m$. Given a subdivision of Δ' into some convex lattice polygons, we extend it up to a subdivision of Δ by cutting the rectangle $\mathrm{conv}\{(0, 0), (0, 1), (a, 0), (a, 1)\}$ with vertical lines through the vertices of the subdivision of Δ' (see Figure 2.11(c)).

A vertex of a convex polygon in $\mathbb{R}^2$ is called h-extremal, if it is a strong extremum of the projection on the horizontal axis. A subdivision of Δ into convex polygons is called **horizontally fibred** if none of the subdivision polygons has h-extremal vertices. If the given subdivision $\mathcal{T} : \Delta = \Delta_1 \cup \cdots \cup \Delta_N$ into convex lattice polygons is not horizontally fibred, then we shall construct a new subdivision $\mathcal{T}^{ref}$ of Δ, which will be horizontally fibred and which we call the **horizontal refinement** of $\mathcal{T}$.

Denote by $\mathrm{Vert}(\mathcal{T})$ the set of vertices of the polygons $\Delta_1, \ldots, \Delta_N$, by $\mathrm{Edges}(\mathcal{T})_0$ the set of the vertical edges of $\Delta_1, \ldots, \Delta_N$ and by $\mathrm{Edges}(\mathcal{T})$ the set of the remaining edges of $\Delta_1, \ldots, \Delta_N$. Consider continuous piecewise-linear functions defined on $[0, a]$ whose graphs are unions of edges belonging to E, and denote by $\mathcal{P}$ the set of the graphs of these functions. For any $v \in \mathrm{Vert}(\mathcal{T})$ and $e \in \mathrm{Edges}(\mathcal{T})$, put

$$\mathcal{P}(v) = \#\{\Gamma \in \mathcal{P} \ : \ v \in \Gamma\}, \quad \mathcal{P}(e) = \#\{\Gamma \in \mathcal{P} \ : \ e \subset \Gamma\} \, .$$

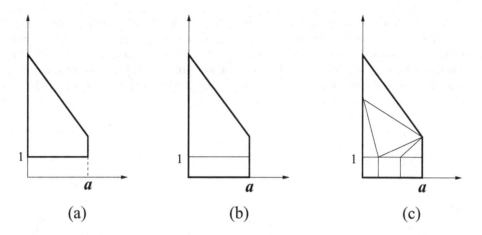

Figure 2.11: Polygons Δ', Δ, and their subdivisions.

Fix a sufficiently small $\varepsilon > 0$ and replace each edge $e \in \mathrm{Edges}(\mathcal{T})$ by the union $\widetilde{e}$ of $2\mathcal{P}(e)$ parallel translates of e in the following way:

- if $e \subset \partial\Delta$, then put

$$\widetilde{e} = \{e,\ tr_\varepsilon(e),\ \ldots,\ tr_\varepsilon^{\mathcal{P}(e)-1}(e)\}\,,$$

  or

$$\widetilde{e} = \{e,\ tr_{-\varepsilon}(e),\ \ldots,\ tr_{-\varepsilon}^{\mathcal{P}(e)-1}(e)\}\,,$$

  where tr_δ is the translation by the vector $(0, \delta)$, and the sign is chosen so that all new edges intersect Δ;

- if $e \not\subset \partial\Delta$, then put

$$\widetilde{e} = \{tr_{-\varepsilon/2}^{2\mathcal{P}(e)-1}(e),\ tr_{-\varepsilon/2}^{2\mathcal{P}(e)-3}(e),\ \ldots,\ tr_{-\varepsilon/2}(e),\ tr_{\varepsilon/2}(e),\ \ldots,\ tr_{\varepsilon/2}^{2\mathcal{P}(e)-1}(e)\}\,.$$

All the endpoints of the edges belonging to $\widetilde{\mathrm{Edges}}(\mathcal{T}) = \bigcup_{e \in \mathrm{Edges}(\mathcal{T})} \widetilde{e}$ and their intersection points lie in small neighborhoods of the vertices of $\mathcal{T}$. For any vertex $v \in \mathrm{Vert}(\mathcal{T})$, we choose such a closed neighborhood in the form of a rectangle R_v^1 with vertical and horizontal sides. Choosing ε smaller if necessary, we can suppose that no edge which belongs to $\widetilde{\mathrm{Edges}}(\mathcal{T})$ intersects the horizontal sides of R_v^1 (see Figure 2.12(a)). Put $R_v^2 = R_v^1 \cap \Delta$.

The intersection points of the edges belonging to $\widetilde{\mathrm{Edges}}(\mathcal{T})$ with the vertical sides of R_v^2 is called **marked points**. If $v = (i, j)$, $a < i < b$, then on the vertical sides of R_v^2 we take the minimal segments s_1, s_2 containing all the marked points, and choose R_v^3 to be the convex hull of the union of s_1, s_2 and the intersection points of R_v^2 with vertical edges of $\mathcal{T}$ adjacent to v (see Figure 2.12(b)). Note that the number of marked points on s_1 and s_2 is the same and equal to $2\mathcal{P}(v)$. Then we subdivide R_v^3 by non-intersecting lines joining the respective marked points on s_1 and s_2 (see Figure 2.12(c)). If the obtained subdivision of R_v^3 contains triangles, then we remove the sides of these triangles lying inside of R_v^3 (see Figure 2.12(d)).

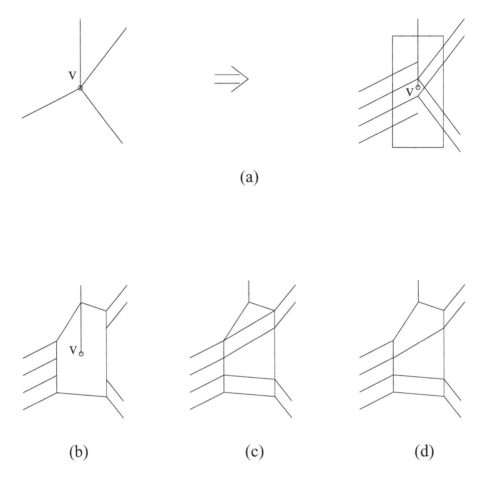

Figure 2.12: Refinement of a subdivision in a neighborhood of a vertex.

If R_v^2 has a vertical edge $s_1 \subset \partial\Delta$ (see Figure 2.13(a)), we take the minimal segment s_2 on the other vertical side of R_v^2, which contains all the intersection points with the edges belonging to $\widetilde{\text{Edges}(\mathcal{T})}$. Then choose R_v^3 to be the convex hull of s_1 and s_2 (see Figure 2.13(b)), and subdivide R_v^3 by horizontal lines through marked points on s_2 (see Figure 2.13(c)). If the obtained subdivision of R_v^3 contains triangles, then we remove the sides of these triangles lying inside of R_v^3

(see Figure 2.13(d)).

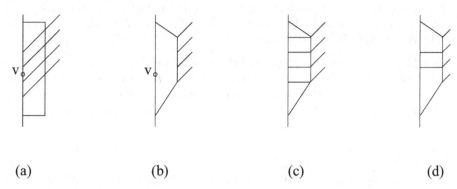

(a)   (b)   (c)   (d)

Figure 2.13: Case when v belongs to a vertical edge of Δ.

Now define the **horizontal refinement** $\mathcal{T}^{ref}$ of Δ into the following convex polygons[3]:

- the polygons of the subdivisions of R_v^3, $v \in \mathrm{Vert}(\mathcal{T})$, introduced above;
- the (closures of the) complements in $\Delta \backslash \bigcup_{v \in \mathrm{Vert}(\mathcal{T})} R_v^3$ to the edges from $\mathrm{Edges}(\mathcal{T})_0 \cup \widetilde{\mathrm{Edges}(\mathcal{T})}$.

The following statement accumulates the key properties of the horizontal refinement.

Lemma 2.26.   (i) *The horizontal refinement is horizontally fibred.*

(ii) *For any $i = 1, \ldots, N$ there is exactly one element Δ_i^{ref} of $\mathcal{T}^{ref}$ such that $\Delta_i^{ref} \subset \Delta_i$ and $\Delta_i \backslash \Delta_i^{ref}$ lies in a small neighborhood of $\partial \Delta_i$. Furthermore, for any vertex v of Δ_i, the polygon Δ_i^{ref} has exactly two vertices in a small neighborhood of v.*

(iii) *For any edge $e \in E$ there are exactly $2\mathcal{P}(e) - 1$ elements of $\mathcal{T}^{ref}$ which are congruent parallelograms, each one with a pair of sides parallel and close to e.*

(iv) *The elements of $\mathcal{T}^{ref}$ which are not mentioned in (i) and (ii) lie in $\bigcup\limits_{v \in \mathrm{Vert}(\mathcal{T})} R_v^3$.*

The proof is straightforward from the construction (see Figures 2.12 and 2.13).

Remark 2.27. According to Remark 2.12, we can modify the moment map and its complexification, spreading the sum in formula (2.2) only to the vertices of the respective polygon. Observe that then the map $\mathbb{C}\mu_{\Delta_i^{ref}}$ is a small equivariant deformation of the map $\mathbb{C}\mu_{\Delta_i}$ for all $i = 1, \ldots, N$.

---

[3]Observe that these polygons are not necessarily lattice polygons.

S^2-bundles over $\mathbb{C}P^1$ associated with horizontally fibred subdivisions. For further consideration we translate Δ and Δ' inside the positive quadrant $\mathbb{R}_+^2$.

Let $\mathcal{S}$ be a horizontally fibred subdivision of Δ, and Λ be the pencil of the (punctured) straight lines $\{y = \mathrm{const}\}$ in $(\mathbb{C}^*)^2$.

Lemma 2.28. (1) *Let δ be a polygon of the subdivision $\mathcal{S}$ which has vertical edges σ_1 and σ_2. Then the charts $\mathbb{C}\,\mathrm{Chart}_\delta(L)$ of lines $L \in \Lambda$ are disjoint and lie in $(\mathrm{Int}(\delta) \cup \mathrm{Int}(\sigma_1) \cup \mathrm{Int}(\sigma_2)) \times (S^1)^2$. Any chart $\mathbb{C}\,\mathrm{Chart}_\delta(L)$ is homeomorphic to a cylinder whose boundary consists of one circle in $\mathrm{Int}(\sigma_1) \times (S^1)^2$ and one circle in $\mathrm{Int}(\sigma_2) \times (S^1)^2$.*

(2) *Let σ be a non-vertical edge of $\mathcal{S}$ with vertices v_1 and v_2. Then the charts $\mathbb{C}\,\mathrm{Chart}_\sigma(L)$, $L \in \Lambda$, are homeomorphic to cylinders whose boundary consists of one circle in $\{v_1\} \times (S^1)^2$ and one circle in $\{v_2\} \times (S^1)^2$. The charts of lines $\{y = c_1\}$, $\{y = c_2\}$ coincide if $c_1/c_2 \in \mathbb{R}$, and are disjoint otherwise.*

(3) *Let δ' and δ'' be two polygons of $\mathcal{S}$ with a common vertical side σ. Then, for any line $L \in \Lambda$, the charts $\mathbb{C}\,\mathrm{Chart}_{\delta'}(L)$ and $\mathbb{C}\,\mathrm{Chart}_{\delta''}(L)$ intersect along their common boundary component in $\sigma \times (S^1)^2$. Similarly, for a pair of non-vertical edges δ' and δ'' of $\mathcal{S}$ with a common vertex v and any line $L \in \Lambda$, the charts $\mathbb{C}\,\mathrm{Chart}_{\delta'}(L)$ and $\mathbb{C}\,\mathrm{Chart}_{\delta''}(L)$ intersect along their common boundary component in $\{v\} \times (S^1)^2$.*

The proof of Lemma 2.28 is straightforward from the definitions and Lemmas 2.16, 2.17, and 2.18.

Let σ_1 and σ_2 be the long and the short vertical sides of Δ, respectively, σ_1' the projection of σ_1 to the vertical axis. This projection naturally extends up to the projection π_1 of $\sigma_1 \times (S^1)^2) \subset \mathbb{C}\Delta$ onto $\mathbb{C}\sigma_1' \simeq \sigma_1 \times S^1$.

Lemma 2.29. *There exists a surjective piece-wise smooth map $\Theta : \mathbb{C}\Delta \to \mathbb{C}\sigma_1'$ such that $\Theta|_{\sigma_1} = \pi_1$, and all fibres of π are unions of charts of lines $L \in \Lambda$ in the complexifications of polygons of $\mathcal{S}$, and are homeomorphic to cylinders with boundary circles in $\sigma_1 \times (S^1)^2$, $\sigma_2 \times (S^1)^2$, respectively.*

The statement immediately follows from Lemma 2.28 and the properties of horizontally fibred subdivisions. The details can be found in the proof of Lemma 4.3 from [26].

The map $\psi_\Delta : \mathbb{C}\Delta \to \mathrm{Tor}_{\mathbb{C}}(\Delta)$ factors $\partial\mathbb{C}\Delta$ by an S^1-action, takes $\mathbb{C}\sigma_1$ to $\mathrm{Tor}_{\mathbb{C}}(\sigma_1) = \mathrm{Tor}_{\mathbb{C}}(\sigma_1') = \mathbb{C}P^1$, and takes each fibre of π_1 into a sphere S^2, contracting the boundary components into points. Observe also that the fibres of π_1 over $\partial\mathbb{C}\sigma_1'$ are identified by ψ_Δ so that the induced fibration $\Theta_\mathcal{S} : \mathrm{Tor}_{\mathbb{C}}(\Delta) = \Sigma_m \to \mathrm{Tor}_{\mathbb{C}}(\sigma_1') = \mathbb{C}P^1$ defines an S^2-bundle with self-intersection $-m$ of the base section $E \overset{\mathrm{def}}{=} \mathrm{Tor}_{\mathbb{C}}(\sigma_1) \subset \mathrm{Tor}_{\mathbb{C}}(\Delta)$. In addition, this fibration commutes with the complex conjugation. Hence there exists an equivariant piece-wise smooth homeomorphism $\mathrm{Tor}_{\mathbb{C}}(\Delta) \to \Sigma_m$ which takes the fibration $\Theta_\mathcal{S}$ to the standard projection π.

Horizontal refinement of M and its deformation into a pseudo-holomorphic curve.
Let us be given a subdivision of Δ' into convex lattice polygons, and a collection
$\{a_{kj} : (k,j) \in \Delta' \cap \mathbb{Z}^2\}$ of numbers such that $a_{kj} \neq 0$ if (k,j) is a vertex of one
of the subdivision polygons. For any subdivision polygon Δ_m, we assume that the
polynomial

$$f_m(x,y) = \sum_{(k,j)\in\Delta_m\cap\mathbb{Z}^2} a_{kj}x^k y^j, \quad m = 1,\ldots,N,$$

is completely non-degenerate. Extend the subdivision of Δ' up to a subdivision $\mathcal{T}$
of Δ as shown in Figure 2.11(c), and, for any subdivision rectangle $\Delta_j \subset \Delta \backslash \Delta'$,
having a common side σ with a subdivision polygon $\Delta_m \subset \Delta'$, put $f_j(x,y) \overset{\text{def}}{=}
F_m^\sigma(x,y)$. Then

$$M \overset{\text{def}}{=} \psi_\Delta \left(\bigcup_{\Delta_m\in\mathcal{T}} \mathbb{C}Ch_{\Delta_i}(f_m) \right) \subset \Sigma_m$$

is a piece-wise smooth surface.

Consider the horizontal refinement $\mathcal{T}^{ref}$ of $\mathcal{T}$. We define a new surface
$M^{ref} \subset \Sigma_m$ as follows. For any $i = 1,\ldots,N$, we take the chart $\mathbb{C}\,\text{Chart}_{\Delta_k^{ref}}(f_k)$,
and for any non-vertical edge σ of Δ_j, $j = 1,\ldots,N$, and any parallelogram
$\delta \in \mathcal{T}^{ref}$ with sides parallel and close to σ, we take the chart $\mathbb{C}\,\text{Chart}_\delta(f_j^\sigma)$.
At last, we define

$$M_{ref} = \psi_\Delta \left(\bigcup_{\Delta_k\in\mathcal{T}} \mathbb{C}Ch_{\Delta_k^{ref}}(f_k) \cup \bigcup_\delta \mathbb{C}\,\text{Chart}_\delta(f_k^\delta) \right) \subset \Sigma_m .$$

Denote by Π the family of (the closures of) the fibres of the fibration on Σ_m,
defined by the subdivision $\mathcal{T}_{ref}$ as described above.

Lemma 2.30. (1) *The set M_{ref} is a piece-wise smooth surface in Σ_m, which is
(equivariantly in the real case) isotopic to M.*

(2) *There is a one-dimensional (over $\mathbb{R}$) set $K \subset \Pi \simeq \mathbb{C}P^1$ such that, any fibre
$\Pi_p \overset{\text{def}}{=} \Theta^{-1}(p) \in \Pi \backslash K$ intersects with M_{ref} exactly at d points, where d is
the length of the projection of Δ on the horizontal coordinate axis. Moreover,
all these intersection points are transverse and positive with respect to the
naturally induced orientations of Π_p and M_{ref}.*

(3) *The surface M_{ref} intersects with E exactly at $l = b - ma$ points. Moreover,
all these intersections are transverse and positive with respect to the natural
orientations of E and M_{ref}.*

Proof. We omit details, and only point out key observations.

The first statement directly follows from the construction, from Remark 2.27.
Then we define K to be the set of fibres in Π, which

- either cross $\{v\} \times (S^1)^2$, where v is a vertex of the subdivision $\mathcal{T}^{ref}$,

- or cross M_{ref} at a point, belonging to $\sigma \times (S^1)^2$, where σ is a vertical edge of the subdivision $\mathcal{T}^{ref}$,

- or are tangent to M_{ref} at a point in $\operatorname{Int} \delta \times (S^1)^2$ for some polygon δ of subdivision $\mathcal{T}^{ref}$.

Now the second statement follows from the fact that the intersection points of Π_p, $p \notin K$, and M_{ref} lie in $\bigcup_\delta \operatorname{Int}(\mathbb{C}\delta)$ with δ running over all polygons of $\mathcal{T}^{ref}$, and hence these intersection points are diffeomorphic (orientation preserving) images of intersection points of $\{y = \operatorname{const}\}$ with $\{f_i = 0\}$ in $(\mathbb{C}^*)^2$. At last, the third statement can be proven in the same way, if we put Δ in the position as in Figure 2.11(b) and then notice that the complexified momentum map smoothly extends to $\mathbb{C}^2$. $\square$

Corollary 2.31. (see [26]) *There exists an almost complex structure on Σ_m and an (equivariant in the real case) isotopy of Σ_m, preserving the lines of the pencil Π, which deforms M into a real pseudo-holomorphic curve.*

Proof. First, we deform M into M_{ref}. Then we deform M_{ref} and the pencil Π so that the intersections of M_{ref} of the fibres of the pencil will satisfy the conditions of the criterion for pseudo-holomorphic curves, formulated in the beginning of Section 2.3.5.

By Lemmas 2.29 and 2.30, the intersection of M_{ref} with the fibred Π_p, $p \notin K$, satisfies these conditions.

Consider now the intersection of M_{ref} with the fibres Π_p, $p \in K$.

If Π_p crosses $\{v\} \times (S^1)^2$ for some vertex v of the subdivision $\mathcal{T}^{ref}$, then by Lemmas 2.28 and 2.29, Π_p lies inside $\bigcup_\sigma \sigma \times (S^1)^2$, where σ runs over some sequence of non-vertical edges of the subdivision $\mathcal{T}^{ref}$. Thus, Π_p crosses M_{ref} at finitely many points lying on $\operatorname{Int}(\sigma) \times (S^1)^2$ for these non-vertical edges σ. Slightly moving Π_p in the pencil Π, we obtain a fibre disjoint with $\bigcup_\sigma \sigma \times (S^1)^2$, which thereby intersects M_{ref} transversally at a smooth points. Hence the initial fibre Π_p crosses M_{ref} at a points with multiplicity $+1$.

Assume that Π_p crosses M_{ref} at a point, belonging to $\sigma \times (S^1)^2$, where σ is a vertical edge of the subdivision $\mathcal{T}^{ref}$. Then $\sigma = \Delta_i^{ref} \cap \Delta_j^{ref}$, where Δ_i, Δ_j are neighboring polygons of the initial subdivision $\mathcal{T}$ of Δ. The intersection $\Pi_p \cap M_{ref}$ contains a circle S^1, lying inside $\sigma \times (S^1)^2$, since the charts of curves in $\sigma \times (S^1)^2$ are products of the charts in $\mathbb{C}\sigma \simeq \sigma \times S^1$ by S^1. We shall deform slightly M_{ref} and the pencil Π in a neighborhood U of that circle S^1 in $\mathbb{C}\Delta$ (here we suppose that the size of U is small, but much larger than that of the small polygons of the subdivision $\mathcal{T}^{ref}$) in order to obtain finite intersections of the surface with the fibres of the pencil in U, which all will be transverse.

Extend the pair of polygons Δ_i, Δ_j up to a convex lattice subdivision $\Delta = \Delta_i \cup \Delta_j \cup \delta_1 \cup \cdots \cup \delta_s$, constructed in any possible way. Let $\nu : \Delta \to \mathbb{R}$ be a convex piecewise-linear function, integral-valued at integral points, whose graph has facets $\Delta_i', \Delta_j', \delta_1', \ldots, \delta_s'$, which respectively project onto the polygons $\Delta_i, \Delta_j, \delta_1, \ldots, \delta_s$.

Introduce the overgraph of ν,

$$\widetilde{\Delta} \stackrel{\text{def}}{=} \{(x, y, \gamma) \in \mathbb{R}^3 \; : \; \nu(x, y) \leq \gamma \leq \nu_0\},$$

with some constant $\nu_0 > \max \nu$. The family of the charts $P_c = \text{Chart}_{\widetilde{\Delta}}(t = c) \subset \mathbb{C}\widetilde{\Delta}$, $c \in (0, \infty)$, uniformly converges as $c \to 0$ to a piece-wise smooth four-manifold with boundary

$$P_0 = \left(\Delta'_i \cup \Delta'_j \cup \delta'_1 \cup \cdots \cup \delta'_s\right) \times (S^1)^2 = \left(\Delta'_i \cup \Delta'_j \cup \delta'_1 \cup \cdots \cup \delta'_s\right) \times (S^1)^2 \times \{1\}$$

$$\subset \left(\Delta'_i \cup \Delta'_j \cup \delta'_1 \cup \cdots \cup \delta'_s\right) \times (S^1)^3 \subset \mathbb{C}\widetilde{\Delta} \,,$$

where the third factor S^1 in the last expression contains the arguments of t. Furthermore, the family P_c, $c \in [0, \infty)$, can be represented by a family of equivariant homeomorphisms $H_c : P_c \to P_0$, $c \in [0, \infty)$, such that $H_0 = \text{Id}$. Introduce the Viro polynomial

$$f_{(t)}(x, y) = \sum_{(i,j) \in \Delta} A_{kj} x^k y^j t^{\nu(k,j)} \,.$$

The intersections $M_c \stackrel{\text{def}}{=} \text{Chart}_{\widetilde{\Delta}}(f_{(t)}) \cap P_c$, $c > 0$, $c \to 0$, converge to a surface (with boundary) $M_0 \subset P_0$, and the intersections $L_{c,\xi} \stackrel{\text{def}}{=} \text{Chart}_{\widetilde{\Delta}}(y = \xi) \cap P_c$, where ξ belongs to a neighborhood U_p of $p \in \mathbb{C}^* \subset E$, converge as $c > 0$, $c \to 0$, to surfaces (with boundary) $L_{0,\xi} \subset P_0$. It can easily be seen that there is a homeomorphism $h : P_0 \to \mathbb{C}\Delta$, which takes $\Delta'_i \times (S^1)^2$ onto $\mathbb{C}\Delta_i$, $\Delta'_j \times (S^1)^2$ onto $\mathbb{C}\Delta_j$, and $\delta'_m \times (S^1)^2$ onto $\mathbb{C}\Delta_m$, $m = 1, \ldots, s$. Observe that the images $h(M_0)$ and $h(L_{0,\xi})$, $\xi \in U_p$, are close in the neighborhood U, chosen above, to $M_{ref} \cap U$ and $\Pi_\xi \cap U$, $\xi \in U_p$, respectively. Then we replace $M_{ref} \cap U$ by $h(H_c(M_c)) \cap U$ and any line $\Pi_\xi \cap U$ by $h(H_c(L_{,\xi})) \cap U$, $\xi \in U_p$, for a sufficiently small fixed $c > 0$. Observe that the intersections of $h(H_c(M_c))$ and $h(H_c(L_{c,\xi}))$ in U are regular (in the sense of the criterion for pseudo-holomorphic curves) for any $\xi \in U_p$. On the other hand there is a (equivariant in the real case) homeomorphism of ∂U onto the respective component of $\partial(\mathbb{C}\Delta \backslash U)$ which takes $h(H_c(M_c)) \cap \partial U$ onto $M_{ref} \cap \partial(\mathbb{C}\Delta \backslash U)$, and takes $h(H_c(L_{,\xi})) \cap \partial U$ onto $\Pi_\xi \cap \partial(\mathbb{C}\Delta \backslash U)$ for all $\xi \in U_p$. Then we glue up U and $\mathbb{C}\Delta \backslash U$ along their boundary using the latter homeomorphism, and obtain a corrected surface M'_{ref} and a corrected pencil Π', which intersect regularly. Now choose a complex structure on Σ_m so that the lines in Π' will be complex straight lines. Then we apply an (equivariant in the real case) isotopy of $\mathbb{C}\Delta$, which preserves the lines of Π' and makes M'_{ref} smooth. This is possible, since along the corners of M'_{ref} its intersections with $L \in \Pi'$ are of multiplicity $+1$.

The proof is completed. $\qquad\qquad\qquad\qquad\qquad\qquad\qquad\qquad\qquad\qquad\quad$ $\square$

2.4 Patchworking of singular algebraic hypersurfaces

The link between patchworking and tropical geometry presented above is not the only one. We intend to demonstrate another link, which appears in the tropical approach to the enumeration of singular algebraic curves.

We start with a modified version of the patchworking construction, which allows one to keep singularities in the patchworking deformation.

An important difference with respect to the original Viro method is that singularities are not stable in general, and thus one has to modify the Viro deformation and impose certain transversality conditions.

We shall treat in more detail a simplified version of patchworking of singular hypersurfaces. Then we shall indicate how to adapt the patchworking theorem to the concrete situation, which appears in the tropical enumeration of nodal curves on toric surfaces.

2.4.1   Initial data

Let us be given:

(1) a classification $\mathcal{S}$ of isolated hypersurface singularities which is finitely determined and are invariant with respect to the $(\mathbb{C}^*)^n$-action, or respectively to the $\mathbb{R}_+^n$-action in the real case;

(2) a convex n-dimensional lattice polytope $\Delta \subset \mathbb{R}^n$ and a subdivision $\Delta = \Delta_1 \cup \Delta_2 \cup \cdots \cup \Delta_N$ into the linearity domains of some convex piecewise-linear function $\nu : \Delta \to \mathbb{R}$ with $\nu(\mathbb{Z}^n) \subset \mathbb{Z}$;

(3) a collection of numbers $a_i \in \mathbb{C}$, or $a_i \in \mathbb{R}$, for $i \in \Delta \cap \mathbb{Z}^n$, such that

   • $a_i \neq 0$ as $a \in \bigcup_{k=1}^N \operatorname{Vert}(\Delta_k)$ and

   • any polynomial

$$f_k(z) = \sum_{i \in \Delta_k \cap \mathbb{Z}^n} a_i z^i$$

   is *peripherally non-degenerate*. That means any truncation f_k^σ on a proper face $\sigma \subsetneq \Delta$ is nonsingular in $(\mathbb{C}^*)^n$, and, for each such polynomial f_k, the set $\operatorname{Sing}(f_k) \stackrel{\text{def}}{=} \operatorname{Sing}(f_k = 0) \cap (\mathbb{C}^*)^n$ should be finite.

Remark 2.32 (**Notation**). For such a polynomial f_k, denote by $\mathcal{S}(f_k) \in \mathbb{Z}^{\mathcal{S}}$ the function

$$\mathcal{S}(f_k) \in \mathbb{Z}^{\mathcal{S}} : \mathcal{S} \longrightarrow \mathbb{Z},$$
$$\mathcal{S} \ni s \longmapsto \#\{z \mid z \text{ is in } \operatorname{Sing}(f_k) \text{ of type } \mathcal{S}\}.$$

2.4.2   Transversality conditions

Introduce the spaces of complex polynomials

$$\mathbb{C}_d[z] = \{\varphi \in \mathbb{C}[z] \mid \deg \varphi \leq d\},$$
$$\mathcal{P}(\Delta) = \{\varphi \in \mathbb{C}[z] \mid \varphi = \sum_{i \in \Delta} \lambda_i z^i\} \, .$$

Let $F \in \mathcal{P}(\Delta)$ have Newton polytope Δ, and let $\partial \Delta_+ \subset \partial \Delta$ be a union of some facets of Δ. Introduce the space of polynomials

$$\mathcal{P}(\Delta, \partial \Delta, F) = \{G \in \mathcal{P}(\Delta) \mid G^\sigma = F^\sigma, \sigma \subset \partial \Delta_+\}$$

(i.e., the space of polynomials from $\mathcal{P}(\Delta)$ coinciding with F on $\partial \Delta_+$).

Definition 2.33. The triple $(\Delta, \partial \Delta_+, F)$ is called *$\mathcal{S}$-transversal* if

- for any $w \in \mathrm{Sing}(F)$ the germ $M_d(w, F) \subset \mathbb{C}_d[z]$ with $d \gg \deg F$ of the family of polynomials having a singular point in a neighborhood of w, which is $\mathcal{S}$-equivalent to $w \in \mathrm{Sing}(F)$, is a smooth complex analytic subset,

- the intersection

$$\bigcap_{w \in \mathrm{Sing}(F)} M_d(w, F) =: M_d(F) \subset \mathbb{C}_d[z]$$

  is transversal in $\mathbb{C}_d[z]$, and

- the intersection

$$M_d(F) \cap \mathcal{P}(\Delta, \partial \Delta_+, F)$$

  is transversal in $\mathbb{C}_d[z]$.

Remark 2.34. Instead of types of isolated singular points one can consider other properties of polynomials (hypersurfaces) which can be localized, are invariant under the torus action, and the corresponding equi-property strata are smooth. Then one can speak of the respective transversality in the sense of Definition 2.33, and prove a series of patchworking theorems similar to that discussed below.

2.4.3 The patchworking theorem

Let G be the adjacency graph of $\Delta_1, \Delta_2, \ldots, \Delta_N$. Now let $\mathcal{G}$ denote the set of oriented graphs with support G having no oriented cycles (see Figure 2.14). For $\Gamma \in \mathcal{G}$ let $\partial \Delta_{k,+}$ be the union of facets of Δ_k, which corresponds to the edges of Γ leaving Δ_k.

Theorem 2.35. *In the above notation, assume that the polynomials $f_1, f_2, \ldots, f_k$ are peripherally non-degenerate, and there is a graph $\Gamma \in \mathcal{G}$ such that all the triples $(\Delta_k, \partial \Delta_{k,+}(\Gamma), f_k)$ are $\mathcal{S}$-transversal. Then there exists a polynomial $f \in \mathcal{P}(\Delta)$ such that*

$$\mathcal{S}(f) = \sum_{k=1}^{N} \mathcal{S}(f_k)$$

and the triple $(\Delta, \emptyset, f)$ is $\mathcal{S}$-transversal.

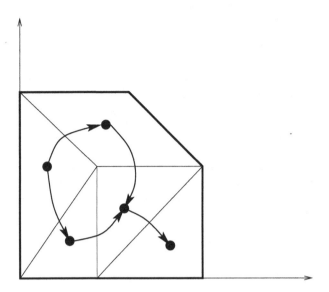

Figure 2.14: An oriented adjacency graph.

Sketch of proof. We look for the desired polynomial f in the family

$$f_{(t)} = \sum_{i \in \Delta \cap \mathbb{Z}^n} A_i(t) t^{\nu(i)} z^i, \qquad t \neq 0,$$

where $A_i(t) = a_i + O(t)$, $i \in \Delta \cap \mathbb{Z}^n$. Here the polynomial $f_{(t)}$ is a modified version of the Viro polynomial. Let $\lambda_k : \mathbb{R}^n \to \mathbb{R}$ be the affine function with $\lambda_k\big|_{\Delta_k} = \nu\big|_{\Delta_k}$ and put $\nu_k = \nu - \lambda_k$ on Δ. Now replacing ν by ν_k we get

$$\begin{aligned} f_{(t),k}(z) &= \sum_{i \in \Delta} A_i(t) t^{\nu_k(i)} z^i \\ &= f_k(z) + O(t) \ . \end{aligned}$$

On the other hand we have

$$f_{(t)}(z) = f_{(t),k}\left(T_k^{(t)}(z)\right) \cdot t^{\alpha_{0,k}} \ ,$$

where

$$\lambda_k(x_1, x_2, \ldots, x_n) = \alpha_{1,k} \cdot x_1 + \alpha_{2,k} \cdot x_2 + \cdots + \alpha_{n,k} \cdot x_n + \alpha_{0,k} \ ,$$

and the transformation $T_k^{(t)}$ is given by

$$T_k^{(t)}(z_1, z_2, \ldots, z_n) = (z_1 \cdot t^{\alpha_{1,k}}, z_2 \cdot t^{\alpha_{2,k}}, \ldots, z_n \cdot t^{\alpha_{n,k}}) \ .$$

Introduce a compact $Q \subset (\mathbb{C}^*)^n$, such that $\mathrm{Int}(Q) \supset \bigcup_{k=1}^N \mathrm{Sing}(f_k)$. Next, choose $t_0 \in \mathbb{R}_+$, such that, for each t with $0 < |t| < t_0$, the compacts

$$\left(T_1^{(t)}\right)^{-1}(Q), \ \left(T_2^{(t)}\right)^{-1}(Q), \ \ldots, \ \left(T_N^{(t)}\right)^{-1}(Q)$$

are pairwise disjoint in $(\mathbb{C}^*)^n$. For $|t|$ small enough the polynomial

$$f_{(t),k}(z) = f_k(t) + O(t) \tag{2.8}$$

is a small deformation of $f_k(z)$ in Q, $k = 1, 2, \ldots, N$.

Now we impose the condition, that (2.8) is an $\mathcal{S}$-equisingular deformation of f_k in Q. Under this condition, $f_{(t)}$ has in

$$\tilde{Q} = \bigcup_{k=1}^N \left(T_k^{(t)}\right)^{-1}(Q)$$

singularities, which are in 1-to-1 $\mathcal{S}$-equivalent correspondence with the disjoint union of the set $\mathrm{Sing}(f_k)$ for $k = 1, 2, \ldots, N$.

Claim: $f_{(t)}$ has no singularities in $(\mathbb{C}^*)^n \setminus \tilde{Q}$.

This, in fact, follows from the peripheral non-degeneracy of $f_1, f_2, \ldots, f_k$, whose complexified charts glue up in a neighborhood of the boundary divisors as in the nonsingular case.

Concluding step: It remains to satisfy the $\mathcal{S}$-equisingular conditions in the deformations

$$f_k \longrightarrow f_{(t),k}, \qquad k = 1, 2, \ldots, N ,$$

by means of a suitable choice of the functions $A_i(t) = a_i + O(t)$. Due to the $\mathcal{S}$-transversality, the germ $M_d(f_k)$ is a smooth complex analytic subset and therefore given by germs of analytic equations

$$\varphi_j^{(k)}(\{B_i \mid |i| \le d\}) = 0, \qquad j = 1, 2, \ldots, d_k.$$

In addition, from the $\mathcal{S}$-transversality of $(\Delta_k, \partial \Delta_{k,+}(\Gamma), f_k)$ it follows that there is a subset

$$\Lambda_k \subset (\Delta_k \setminus \Delta_{k,+}(\Gamma)) \cap \mathbb{Z}^n, \qquad |\Lambda_k| = d_k,$$

such that $\det \mathcal{D}_k \ne 0$ at $\hat{B}_i$, where $\mathcal{D}_k$ is given by

$$\mathcal{D}_k = \frac{\partial \left\{ \varphi_j^{(k)}(\{B_i \mid i \in \Delta\}) \mid j = 1, 2, \ldots, d_k \right\}}{\partial \{B_i \mid i \in \Lambda_k\}}$$

$$\text{as } \hat{B}_i = \begin{cases} a_i, & i \in \Delta_k, \\ 0, & \text{otherwise.} \end{cases}$$

Next we extend a partial order on $\Delta_1, \Delta_2, \ldots, \Delta_N$, defined by the graph Γ, up to a linear order. Then determine $A_i(t) = a_i + O(t)$ from the system of equations

$$\varphi_j^{(k)}\left(A_i \cdot t^{\nu_k(i)} \mid i \in \Delta\right) = 0,$$

$$j = 1, 2, \ldots, d_k, \quad k = 1, 2, \ldots, N$$

which has the solution $\hat{A}_i = \hat{B}_i$, $i \in \Delta \cap \mathbb{Z}^n$, as $t = 0$. Therefore it is soluble for small values of t by the implicit function theorem, since the sets $\Lambda_1, \Lambda_2, \ldots, \Lambda_N \subset \Delta \cap \mathbb{Z}^n$ are disjoint by construction, and

$$\det\left(\frac{\partial\left\{\varphi_j^{(k)}(A_i \cdot t^{\nu_k(i)} \mid i \in \Delta) \;\middle|\; \begin{array}{l} j = 1, 2, \ldots, d_k \\ k = 1, 2, \ldots, N \end{array}\right\}}{\partial\{A_i \mid i \in \Lambda_1 \cup \Lambda_2 \cup \cdots \cup \Lambda_N\}}\right) \neq 0$$

as $t = 0$. The latter can easily be extracted from the fact that the above matrix takes a block-triangular form as $t = 0$ with the blocks $\mathcal{D}_1, \mathcal{D}_2, \ldots, \mathcal{D}_N$ on the diagonal.

The theorem follows.   □

Remark 2.36. The core of the above proof is as follows. The linear space of polynomials with Newton polygon Δ splits into the direct sum of subspaces generated by the monomials z^i, $i \in (\Delta_k \backslash \partial \Delta_{k,+}(\Gamma)) \cap \mathbb{Z}^n$, for $k = 1, \ldots, N$. The transversality conditions mean that one can preserve the singularities, coming from $(\mathbb{C}^*)^n \subset \mathrm{Tor}(\Delta_k)$, by variation of the coefficients of the monomials z^i, $i \in (\Delta_k \backslash \partial \Delta_{k,+}(\Gamma)) \cap \mathbb{Z}^n$, which compensate the distortion caused by the remaining monomials. In turn, the total order of these polynomial subspaces, induced by the graph Γ, prevents cycles in these mutual compensations.

2.4.4   Some $\mathcal{S}$-transversality criteria

Let $n = 2$, i.e., $f_1, f_2, \ldots, f_N$ define curves on toric surfaces.

Claim: There exists a non-negative integer topological invariant $b(w)$ of isolated planar curve singular points w such that, if f_k is irreducible and

$$\sum_{w \in \mathrm{Sing}(f_k)} b(w) < \sum_{\sigma \not\subset \partial \Delta_{k,+}} \mathrm{length}(\sigma) ,$$

then the triple $(\Delta_k, \partial \Delta_{k,+}(\Gamma), f_k)$ is $\mathcal{S}$-transversal.

The precise definition of the invariant $b(w)$ can be found, for example, in [56]. Here we simply recall that

- if w is a node, then $b(w) = 0$,

- if w is a cusp, then $b(w) = 1$,

- if the singularity of w is defined to have tangency of order m to a given line $L \supset \{w\}$ at w, then $b(w) = m$,

- if the singularity of w is defined to have tangency of order m to a given line $L \supset \{w\}$ at a point near to w, then $b(w) = m - 1$.

Example 2.37. If under the hypotheses of the theorem on patchworking of singular hypersurfaces, $n = 2$, the curves $\overline{\{f_k = 0\}} \subset \mathrm{Tor}(\Delta_k)$, $k = 1, \ldots, N$, are irreducible and have only ordinary nodes as singularities, then there is an oriented graph $\Gamma \in \mathcal{G}$ such that all the triples $(\Delta_k, \partial\Delta_{k,+}(\Gamma), f_k)$ are transversal. Indeed, orient the arcs of Γ (supposed to be orthogonal to the dual edges of $\Delta_1, \Delta_2, \ldots, \Delta_N$) so that they form angles in the interval $(-\pi/2, \pi/2]$ with the horizontal axis. Then

$$\partial\Delta_{k,+}(\Gamma) \neq \partial\Delta_k, \qquad k = 1, 2, \ldots, N$$

and the above criterion implies transversality.

Remark 2.38. We should like to point out that the patchworking of pseudo-holomorphic curves, treated in Section 2.3.5, can naturally be extended to the case of patchworking of curves having isolated singularities in $(\mathbb{C}^*)^2$, where as a result we obtain pseudo-holomorphic curves with locally analytic isolated singularities. Moreover, this construction does not require any transversality conditions and can always be performed with any number and types of singularities.

2.5   Tropicalization and patchworking in the enumeration of nodal curves

The aim of this section is to demonstrate how the patchworking construction and the tropicalization procedure (an inverse in a sense operation) apply to the enumeration of real and complex nodal curves on toric surfaces. In particular, we prove Theorem 3.6 in Chapter 3 (Mikhalkin's correspondence theorem for the plane) and Theorem 3.10 in Chapter 3 (the tropical formula for the Welschinger invariants of the plane). We also explain Mikhalkin's algorithm which reduces the count of tropical curves to enumeration of lattice paths in a given Newton polygon (see Section 3.6 in Chapter 3).

After basic definitions concerning plane tropical curves, we state the enumerative problem and present the tropical formulas for the Gromov–Witten and Welschinger invariants. Then we describe the tropical limits of nodal curves in a toric surface over a non-Archimedean field (Lemma 2.46 in Section 2.5.5), afterwards refine the tropical limits, adding an extra piece of information. In the next step we describe how to restore the tropical curves and appropriate tropical limits out of the given configuration of points in the toric surface. Finally, the patchworking Theorem 2.51 (Section 2.5.10) and the refined condition to pass through the given configuration of points on the toric surface (Section 2.5.9) provide us with the required number of algebraic curves, projecting to the same tropical curve.

We end up with an explanation of how to obtain the tropical formula for the Welschinger invariants of toric Del Pezzo surfaces with the standard real structure (Section 2.5.11).

2.5.1 Plane tropical curves

A plane tropical curve T with Newton polygon $\Delta \subset \mathbb{R}^2$ can be defined as the corner locus of a tropical polynomial

$$N(x) = \max_{i \in \Delta \cap \mathbb{Z}^2} (xi + c_i), \quad x \in \mathbb{R}^2 , \tag{2.9}$$

which is a graph whose edges are equipped with positive integral weights: the weight $w(e)$ of an edge e, on which two linear affine functions $xi_1 + c_{i_1}$ and $xi_2 + c_{i_2}$ in the right-hand side of (2.9) coincide, is the greatest common divisor of the coordinates of the vector $i_1 - i_2$. A tropical curve T satisfies the following balancing condition at each of its vertices v:

$$\sum_{v \in e} w(e) \cdot u(e, v) = 0 , \tag{2.10}$$

where e runs over all edges of T adjacent to v, and $u(e, v)$ is the primitive integral vector along e oriented out of v.

The Legendre transform takes the tropical polynomial N to a convex piece-wise-linear function $\nu : \Delta \to \mathbb{R}$, whose linearity domains are convex lattice polygons, forming a subdivision $\Delta = \Delta_1 \cup \cdots \cup \Delta_N$. This subdivision S is dual to the tropical curve T in the following sense:

- the components of $\mathbb{R}^2 \backslash T$ are in 1-to-1 correspondence with the vertices of the subdivision S,

- the edges of T are in 1-to-1 correspondence with the edges of the subdivision S so that an edge e of T is dual to an orthogonal edge of the subdivision S, having the lattice length equal to $w(e)$,

- the vertices of T are in 1-to-1 correspondence with the polygons $\Delta_1, \ldots, \Delta_N$ so that the valency of a vertex of T is equal to the number of sides of the dual polygon.

A plane tropical curve with the Newton polygon Δ is called **nodal**, if its dual subdivision S of Δ consists of only triangles and parallelograms. A nodal tropical curve is called **simple** if, in addition, all the integral points on $\partial \Delta$ are vertices of S. The Mikhalkin multiplicity $\mu(T)$ of a simple tropical curve T is the product of areas of all the triangles in S (we normalize the area in such a way that the area of a triangle whose only integer points are the vertices is equal to 1). The Welschinger multiplicity $\mathcal{W}(T)$ of a simple tropical curve is equal to 0, if T has at least one edge of even weight (equivalently, if the dual subdivision S contains an edge of even length), and is equal to $(-1)^{s(T)}$, if all the edges of T have odd weights, where $s(T)$ is the total number of integral points in the interior of all the triangles of the subdivision S.

A tropical curve is called irreducible if it is not the union of two proper tropical subcurves.

Tropical curves with given Newton polygon Δ, which are dual to the same convex lattice subdivision S of Δ, are parameterized by a convex polyhedron $\mathcal{T}(\Delta, S)$, whose dimension is called the **rank** of the corresponding tropical curves. For example, for nodal curves we have

Lemma 2.39. *The rank of a nodal tropical curve T is equal to the number of the vertices of the dual subdivision S diminished by 1 and by the number of parallelograms in S.*

The proof is left to the reader as exercise.

In general, one can obtain the following estimate (see [59], Lemma 2.2).

Lemma 2.40. *For an arbitrary plane tropical curve T,*

$$\mathrm{rank}_{exp}(T) \overset{def}{=} |\mathrm{Vert}(S)| - 1 - \sum_{k=1}^{N}(|\mathrm{Vert}(\Delta_k)| - 3) \leq \mathrm{rank}(T) \ .$$

That is

$$\mathrm{def}(T) \overset{def}{=} \mathrm{rank}(T) - \mathrm{rank}_{exp}(T)$$

is always non-negative. Furthermore, if T is nodal, then $\mathrm{def}(T) = 0$, and if T is not nodal, then

$$2 \cdot \mathrm{def}(T) \leq \sum_{m \geq 2}((2m-3)N_{2m} - N'_{2m}) + (2m-2)N_{2m+1} - 1 \ , \tag{2.11}$$

where $N_s, s \in \mathbb{N}$, means the number of s-valent vertices of T and N'_{2m} the number of $2m$-valent vertices of T, which locally are intersections of m straight lines.

Definition 2.41. We say that the distinct points $x_1, \ldots, x_\zeta \in \mathbb{Q}^2$ are in (Δ, S)-general position, if the condition to pass through $x_1, \ldots, x_\zeta$ so that $k \leq \zeta$ given points are vertices of a tropical curve, cuts out of $\mathcal{T}(\Delta, S)$ either the empty set, or a polyhedron of codimension $(\zeta - k) + 2k = \zeta + k$. We say that the distinct points $x_1, \ldots, x_\zeta$ are in Δ-general position, if they are (Δ, S)-general for all convex lattice subdivisions S of Δ.

Lemma 2.42. *For any given convex lattice polygon Δ, the set of Δ-general configurations $x_1, \ldots, x_\zeta$ is dense in the set of all ζ-tuples in $\mathbb{Q}^2$.*

This statement is rather clear, since the set of Δ-general configurations is the complement of finitely many hyperplanes in $(\mathbb{Q}^2)^\zeta$ (the finiteness here comes from the finiteness of the set of possible convex lattice subdivisons of the given Δ and from the finiteness of the set of slopes of lattice segments in Δ).

2.5.2 Algebraic enumerative problem and its tropical analogue

Let $\Delta \subset \mathbb{R}^2$ be a non-degenerate convex lattice polygon, $\mathcal{L}_\Delta$ the tautological line bundle on $\mathrm{Tor}(\Delta)$ (i.e., generated by monomials $z^k w^j$, $(k,j) \in \Delta \cap \mathbb{Z}^2$). Let n be a non-negative integer smaller than or equal to $|\mathrm{Int}(\Delta \cap \mathbb{Z}^2)|$. Denote by $\mathrm{Sev}_n(\Delta)$ the set of curves C in the linear system $|\mathcal{L}_\Delta|$, which have n nodes as their only singularities. It is a smooth quasiprojective variety of pure dimension

$$\zeta \overset{\text{def}}{=} \dim \mathrm{Sev}_n(\Delta) = \dim |\mathcal{L}_\Delta| - n = |\Delta \cap \mathbb{Z}^2| - 1 - n.$$

Furthermore, it contains the subset $\mathrm{Sev}_n^{irr}(\Delta)$ of irreducible curves, which either is empty, or has pure dimension ζ. The enumerative geometry problems are formulated as

Question: *What are* $\deg \mathrm{Sev}_n(\Delta)$ *and* $\deg \mathrm{Sev}_n^{irr}(\Delta)$ *?*

These numbers are sometimes called the **relative Gromov–Witten invariants** of the toric surface $\mathrm{Tor}(\Delta)$.

We want to translate the above question into a tropical geometry problem.

For this purpose, we restate it over the field $\mathbb{K} = \bigcup_{m \geq 1} \mathbb{C}\{\{t^{1/m}\}\}$ of locally convergent complex Puiseux series. This is an algebraically closed field of characteristic zero with a non-Archimedean valuation

$$\mathrm{val}\left(\sum_{\tau \in \mathbb{Q}} a_\tau \cdot t^\tau \right) = -\min\{\tau \mid a_\tau \neq 0\} .$$

Geometrically, the degrees $\deg \mathrm{Sev}_n(\Delta)$ and $\deg \mathrm{Sev}_n^{irr}(\Delta)$ can be viewed as follows. Fix ζ generic points $p_1, p_2, \ldots, p_\zeta \in (\mathbb{K}^*)^2 \subset \mathrm{Tor}_K(\Delta)$. Then $\deg \mathrm{Sev}_n(\Delta)$ (resp., $\deg \mathrm{Sev}_n^{irr}(\Delta)$) is equal to the number of curves in the set

$$\mathrm{Sev}_n(\Delta; p_1, \ldots, p_\zeta) := \{C \in \mathrm{Sev}_n(\Delta) : p_1, \ldots, p_\zeta \in C\}$$

(resp., in the set

$$\mathrm{Sev}_n^{irr}(\Delta; p_1, \ldots, p_\zeta) := \{C \in \mathrm{Sev}_n^{irr}(\Delta) : p_1, \ldots, p_\zeta \in C\}).$$

Assuming that

$$x_1 = \mathrm{val}(p_1), \quad x_2 = \mathrm{val}(p_2), \quad \ldots, \quad x_\zeta = \mathrm{val}(p_\zeta) \in \mathbb{Q}^2 \subset \mathbb{R}^2$$

are distinct Δ-generic points, where $\mathrm{val}(*)$ means the coordinate-wise valuation, we consider the tropical curves A_C with Newton polygon Δ, supported at the non-Archimedean amoebas of the curves $C \in \mathrm{Sev}_n(\Delta; p_1, \ldots, p_\zeta)$ (or, $C \in \mathrm{Sev}_n^{irr}(\Delta; p_1, \ldots, p_\zeta)$), i.e., the closures of the images $\mathrm{val}(C \cap (\mathbb{K}^*)^2)$ in $\mathbb{R}^2$, which then pass through the points $x_1, \ldots, x_\zeta$. Thus, we face the two tasks:

- describe and enumerate the tropical curves A_C, which are projections of the curves $C \in \mathrm{Sev}_n(\Delta; p_1, \ldots, p_\zeta)$ (resp., $C \in \mathrm{Sev}_n^{irr}(\Delta; p_1, \ldots, p_\zeta)$), and

- find how many curves $C' \in \mathrm{Sev}_n(\Delta; p_1, \ldots, p_\zeta)$ (resp., $C' \in \mathrm{Sev}_n^{irr}(\Delta; p_1, \ldots, p_\zeta)$) are projected to each of the above tropical curves.

The answers (which are traditionally called the **correspondence statements**) are given in Theorem 2.43 below.

To fulfill the first task, we consider the curves over $\mathbb{K}$ as one-parametric families of curves over $\mathbb{C}$, and define their tropical limits, which consist of a tropical part, i.e., the respective tropical curve, and an algebraic part, a collection of certain complex algebraic curves, and then refine it. In its turn, the patchworking technique allows one to show that each combination of a tropical curve and a collection of complex curves, satisfying certain conditions, admits a one-parameter deformation, which can be viewed as an algebraic curve over the field $\mathbb{K}$.

The above enumerative question has an important real counterpart, the computation of the Welschinger invariants. This problem is discussed in Chapter 3 for the case of the projective plane. We present here a more general result. Let Σ be a toric Del Pezzo surface, equipped with the standard real structure, i.e., Σ is either the plane $\mathbb{P}^2$, or the quadric $(\mathbb{P}^1)^2$, or $\mathbb{P}^2_k$, the plane blown up at $k = 1, 2, 3$ real points (not lying on a line as $k = 3$). One can represent Σ as $\mathrm{Tor}(\Delta)$, where Δ is one of the lattice polygons with the side slopes 0, -1, ∞, shown in Figure 2.15, and the complex conjugation acts trivially on the standard basis $x^k y^j$, $(k, j) \in \Delta$, of the linear system $|\mathcal{L}_\Delta|$.

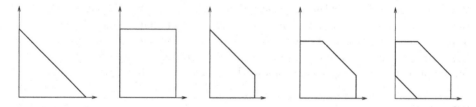

Figure 2.15: Polygons associated with toric Del Pezzo surfaces.

Given a generic configuration of $\zeta = |\partial\Delta \cap \mathbb{Z}^2| - 1$ real points in $\mathrm{Tor}(\Delta)$, there are only finitely many real rational curves $C \in |\mathcal{L}_\Delta|$, passing through the given points. Summing up the weights $(-1)^{s(C)}$ over the latter set, where $s(C)$ is the number of real solitary nodes of C, we obtain a number $W_0(\Delta)$, which does not depend on the choice of the generic configuration of fixed points (Welschinger's theorem [73, 74]) and is called the **Welschinger invariant** of the surface $\mathrm{Tor}(\Delta)$ associated with the linear system $|\mathcal{L}_\Delta|$ (cf. Section 3.7 in Chapter 3). The tropical approach applies to the computation of Welschinger invariants as well, due to the fact that the tropicalization procedure and the patchworking construction respect the real structure, i.e., from real data they produce real objects as results. In the next section we present Theorem 2.44 with the tropical formula for the Welschinger invariants of toric Del Pezzo surfaces, equipped with the standard real structure (in particular, for the plane, as considered in Theorem 3.10, Chapter 3).

2.5.3   Tropical formulas for the Gromov–Witten and Welschinger invariants

Theorem 2.43. *Given a nondegenerate lattice polygon $\Delta \subset \mathbb{R}^2$ and a non-negative integer $n \leq |\operatorname{Int}(\Delta) \cap \mathbb{Z}^2|$, we have*

$$\deg \operatorname{Sev}_n(\Delta) = \sum_{T \in \mathcal{T}_{\Delta,\zeta}(\mathcal{U})} \mu(T), \quad \deg \operatorname{Sev}_n^{irr}(\Delta) = \sum_{T \in \mathcal{T}_{\Delta,\zeta}^{irr}(\mathcal{U})} \mu(T) ,$$

where $\mathcal{U} \subset \mathbb{Q}^2$ is an arbitrary Δ-general configuration of $\zeta = |\Delta \cap \mathbb{Z}^2| - 1 - n$ distinct points, $\mathcal{T}_{\Delta,\zeta}(\mathcal{U})$ (resp., $\mathcal{T}_{\Delta,\zeta}^{irr}(\mathcal{U})$) is the set of all (resp., only irreducible) simple tropical curves with Newton polygon Δ of rank ζ.

Theorem 2.44. *Let Δ be one of the polygons that are shown in Figure 2.15, $\zeta = |\partial\Delta \cap \mathbb{Z}^2| - 1$, and let $\mathcal{U} \subset \mathbb{Q}^2$ be a Δ-general configuration of ζ distinct points. Then*

$$W_0(\Delta) = \sum_{T \in \mathcal{T}_{\Delta,\zeta}^{irr}(\mathcal{U})} \mathcal{W}(T) .$$

We remark that in the computation of Welschinger invariants, corresponding to configurations with imaginary points (cf. Section 3.12 in Chapter 3), or Welschinger invariants of the toric Del Pezzo surfaces with non-standard real structures, one faces additional problems caused by the fact that a generic configuration of points in $\operatorname{Tor}(\Delta)$ projects by the valuation to a non-generic configuration (i.e., the projected configuration contains coinciding points or is symmetric). However the tropical technique gives answers in these cases as well [60, 61].

2.5.4   Tropical limit

Let a polynomial

$$f(z, w) = \sum_{(i,j) \in \Delta} a_{kj}(t) \cdot z^k w^j \in \mathbb{K}[z, w]$$

define a curve $C \subset \operatorname{Tor}_\mathbb{K}(\Delta)$ with only isolated singularities. Changing the parameter $t \mapsto t^M$, we can make all the exponents of t in $a_{kj}(t)$ with $(k, j) \in \Delta$ integral, and thus, we obtain an analytic family $C^{(t)}$ of complex curves in $\operatorname{Tor}(\Delta)$ for $\{0 < |t| < \delta\} = \mathcal{D}\backslash\{0\}$.

Lemma 2.45. *For small δ, the family $C^{(t)}$ is equisingular, and the topological types of singularities of $C^{(t)}$ are in 1-to-1 correspondence with topological types of singularities of $C = \{f = 0\} \subset \operatorname{Tor}_\mathbb{K}(\Delta)$.*

We extend the families

$$
\begin{array}{ccc}
\mathrm{Tor}(\Delta) \times (\mathcal{D}\backslash\{0\}) & \longleftarrow & C \\
\downarrow & & \downarrow \\
\mathcal{D}\backslash\{0\} & \overset{=}{\longleftarrow} & \mathcal{D}\backslash\{0\}
\end{array}
$$

to the origin as follows. Take the tropical polynomial

$$
N_f(x) = \max_{i \in \Delta \cap \mathbb{Z}^2} \left(xi + \mathrm{val}(a_i) \right), \quad x \in \mathbb{R}^2 ,
$$

and consider the function $\nu = \nu_f : \Delta \to \mathbb{R}$, Legendre dual to N_f. It is convex, piecewise-linear, and its linearity domains form a subdivision of Δ into convex lattice polygons: $\Delta = \Delta_1 \cup \cdots \cup \Delta_N$. Furthermore, we can write

$$
f(z, w) = \sum_{(i,j) \in \Delta} (a_{kj}^{(0)} + O(t)) t^{\nu(k,j)} z^k w^j ,
$$

where $a_{kj}^{(0)} \in \mathbb{C}$ are non-zero at least for all the vertices of $\Delta_1, \ldots, \Delta_N$. We define the *tropicalization* (tropical limit) of f as a pair consisting of

(1) the tropical curve T_f defined by the tropical polynomial N_f, or, equivalently, the function ν_f and the corresponding subdivision $S_f : \Delta = \Delta_1 \cup \Delta_2 \cup \cdots \cup \Delta_N$ and

(2) the collection of curves $C_m \subset \mathrm{Tor}(\Delta_m)$ for $m = 1, 2, \ldots, N$ defined by

$$
f_m(z, w) = \sum_{(i,j) \in \Delta_m} a_{kj}^{(0)} z^k w^j .
$$

The family $\mathrm{Tor}(\Delta) \times (\mathcal{D}\backslash\{0\}) \to \mathcal{D}\backslash\{0\}$ extends up to $\mathrm{Tor}(\widetilde{\Delta}) \to \mathbb{C}$ with $\widetilde{\Delta}$ being the overgraph of ν_f [4] (cf. with the same extension in the proof of Viro's patchworking theorem, Section 2.3.2), and the family $C^{(t)}$, $t \neq 0$, is completed at $t = 0$ by

$$
C^{(0)} = \bigcup_{m=1}^{N} C_m \subset \bigcup_{m=1}^{N} \mathrm{Tor}(\Delta_m) \subset \mathrm{Tor}\left(\widetilde{\Delta}\right) .
$$

2.5.5 Tropicalization of nodal curves

Lemma 2.46. *Let the points $x_1 = \mathrm{val}(p_1)$, $\ldots$, $x_\zeta = \mathrm{val}(p_\zeta)$ be Δ-generic. Then the tropical limit of a curve $C \in \mathrm{Sev}_n(\Delta)$ (or $C \in \mathrm{Sev}_n^{irr}(\Delta)$), passing through $p_1, p_2, \ldots, p_\zeta$ consists of*

---

[4]We suppose here that $\nu_f(\mathbb{Z}^2) \subset \mathbb{Z}$.

(1) *a simple (resp., simple irreducible) tropical curve T with Newton polygon Δ of rank $A = \zeta$, and*

(2) *a collection of curves $C_m \subset \mathrm{Tor}(\Delta_m)$ for $m = 1, 2, \ldots, N$ such that the following holds:*

   - *if Δ_m is a triangle, then $C_m \in |\mathcal{L}_{\Delta_k}|$ is an irreducible rational nodal curve crossing $\mathrm{Tor}(\partial\Delta_m) \overset{def}{=} \bigcup_{\sigma \subset \mathrm{Tor}(\Delta_m)} \mathrm{Tor}(\sigma)$ at precisely three points, where it is unibranch;*

   - *if Δ_m is a parallelogram, then C_m is given by $f_m \in \mathcal{P}(\Delta_m)$, which splits into a product of a monomial and binomials.*

Proof of Lemma 2.46. Our strategy is to estimate $\hat{\chi}(C^{(t)})$, $t \neq 0$, from above and below, and then to extract the required statements from the comparison of the obtained estimates. Here $\hat{\chi}(C^{(t)})$ is meant to be the Euler characteristic of the normalization of $C^{(t)}$.

Upper bound. Take a small regular neighborhood U of $\bigcup_{k=1}^{N} \mathrm{Tor}(\partial\Delta_k)$ in $\mathrm{Tor}(\tilde{\Delta})$. Then the Euler characteristic can be calculated via

$$\hat{\chi}(C^{(t)}) = \hat{\chi}(C^{(t)} \cap U) + \hat{\chi}(C^{(t)} \backslash U) \ .$$

Now we claim that

$$\hat{\chi}(C^{(t)} \cap U) \leq \mathrm{Br}(C^{(0)}, \partial\Delta) \ , \tag{2.12}$$

where $\mathrm{Br}(C^{(0)}, \partial\Delta)$ is the number of local branches of the curves $C_1, C_2, \ldots, C_N$ centered along $\bigcup_{\sigma \subset \partial\Delta} \mathrm{Tor}(\sigma) \subset \mathrm{Tor}(\tilde{\Delta})$ counted with their multiplicities, if they are not reduced.

This actually follows from the fact that a local branch of C_m centered along a divisor $\mathrm{Tor}(\sigma)$, $\sigma = \Delta_m \cap \Delta_k$, and which topologically is a disc, cannot stay as a disc in the deformation. Therefore it must join with another branch by a handle.

Next we have

$$\hat{\chi}(C^{(t)} \backslash U) \leq \sum_{k=1}^{N} \sum_{j} \hat{\chi}(C_{kj} \backslash U)$$

$$\leq \sum_{k=1}^{N} \sum_{j} (\, 2 - \mathrm{Br}(C_{kj}, \partial\Delta_k) \,)$$

$$\leq - \sum_{m \geq 2} (N_{2m-1} + N_{2m} - N'_{2m}) \ .$$

Here C_{kj} denotes all components of C_k counted with their multiplicities. Furthermore, the equality in the latter relation yields that all the components C_{kj} are rational, and, moreover, it follows that they do not glue up with each other by handles in $\mathrm{Tor}(\tilde{\Delta}) \backslash U$ along the deformation $C^{(t)}, t \in \mathcal{D}$. Moreover, in the last relation, the equality means that

- for any $2m$-gon, whose opposite edges are parallel, f_k splits into a product of binomials, i.e., $\mathrm{Br}(C_{kj}, \partial\Delta_k) = 2$ for all C_{kj}, and,

- for any other polygon Δ_k precisely one component C_{kj} (counting with multiplicity) has $\mathrm{Br}(C_{kj}, \partial\Delta_k) = 3$ and for all other components it holds that $\mathrm{Br}(C_{kj}, \partial\Delta_k) = 2$, i.e., they are defined by binomials.

So, finally, we get the following upper bound:

$$\hat{\chi}(C^{(t)}) \leq \mathrm{Br}(C^{(0)}, \partial\Delta) - \sum_{m \geq 2} (N_{2m-1} + N_{2m} - N'_{2m}) \ .$$

Lower bound. Since

$$\zeta = |\partial\Delta \cap \mathbb{Z}^2| + g - 1 \ ,$$

we obtain (here S denotes the subdivision of Δ dual to T):

$$\begin{aligned}
\hat{\chi}(C^{(t)}) &= 2 - g(C^{(t)}) = 2 - 2(\zeta + 1 - |\partial\Delta \cap \mathbb{Z}^2|) \\
&= 2|\partial\Delta \cap \mathbb{Z}^2| - 2\zeta \geq 2|\partial\Delta \cap \mathbb{Z}^2| - 2\,\mathrm{rank}(T) \\
&= 2|\partial\Delta \cap \mathbb{Z}^2| - 2\,\mathrm{rank}_{\exp}(T) - 2\,\mathrm{def}(T) \\
&= 2|\partial\Delta \cap \mathbb{Z}^2| - 2|\,\mathrm{Vert}(S)| + 2 + 2\sum_{k=1}^{N}(|\,\mathrm{Vert}(\Delta_k)| - 3) - 2\,\mathrm{def}(T) \\
&= 2|\partial\Delta \cap \mathbb{Z}^2| - 2|\mathrm{Vert}(S)| + 2 \\
&\quad + 2|\mathrm{Vert}(S) \cap \partial\Delta| + 4|\mathrm{Edges}(S)| - 6N - 2\,\mathrm{def}(T) \\
&= 2(|\partial\Delta \cap \mathbb{Z}^2| - |\mathrm{Vert}(S) \cap \partial\Delta|) + |\mathrm{Vert}(S) \cap \partial\Delta| \\
&\quad + \sum_{m \geq 3}(m - 4)N_m - 2\,\mathrm{def}(T).
\end{aligned}$$

Combining the upper and lower bound of $\hat{\chi}(C^{(t)})$, we conclude

$$\begin{aligned}
2|\partial\Delta \cap \mathbb{Z}^2| &- |\mathrm{Vert}(S) \cap \partial\Delta| - \mathrm{Br}(C^{(0)}, \partial\Delta) \\
&\leq 2\,\mathrm{def}(T) - \sum_{m \geq 2}((2m - 3)N_{2m} - N'_{2m}) - \sum_{m \geq 2}(2m - 2)N_{2m+1} \ ,
\end{aligned}$$

which in view of the upper bound (2.11) to $\mathrm{def}(T)$ and the equality conditions implies the lemma, and, in addition, that $\mathrm{rank}_{\exp}(T) = \mathrm{rank}(T) = \zeta$ and $\mathrm{def}(T) = 0$.

One has only to confirm the irreducibility of T. Indeed, if $T = T_1 \cup T_2$, then T_1 and T_2 cross only at four-valent vertices of T. Then the components of $C^{(0)}$ naturally form two subsets and the intersections of the components from different sets fall into the toric part of $\mathrm{Tor}(\Delta_k)$ for some parallelograms Δ_k. These intersecting components of $C^{(0)}$ do not glue up in the deformation $C^{(t)}$ and hence $C^{(t)}$ turns out to be reducible — a contradiction. $\qquad\square$

Geometric point of view on the proof of Lemma 2.46. We recall that the genus of a tropical curve T is defined as the minimal $b_1(\Gamma)$ over all graphs Γ, parameterizing T (see details in Section 1.6, Chapter 1, or in [41, 42, 45, 46]). In particular, for a nodal tropical curve T, the suitable parameterizing graph Γ can be obtained by resolving the four-valent vertices of T (i.e., by separating the two smooth local branches of T, intersecting at that vertex). The following statement is a kind of a tropical Riemann–Roch theorem.

Lemma 2.47. *For a nodal tropical curve T,*

$$\mathrm{rank}(T) = |\mathrm{Ends}(T)| + g(T) - 1 \;, \tag{2.13}$$

where $\mathrm{Ends}(T)$ is the set of infinite edges of T, and $g(T)$ is its genus.

The proof is left to the reader as exercise. We only show that formula (2.13) agrees with the statement of Lemma 2.39. Indeed, compactify $\mathbb{R}^2$ up to $S^2 = \mathbb{R}^2 \cup \{\infty\}$ and add the point ∞ to T as well. Then the number of independent cycles in $T \cup \{\infty\}$ is $|\pi_0(\mathbb{R}^2 \backslash T)| - 1 = |\mathrm{Vert}(S)| - 1$, where S is the dual subdivision of the Newton polygon. Removing the added point ∞ we break $|\mathrm{Ends}(T)| - 1$ cycles. Resolving the four-valent vertices of T, we remove each time another cycle. Counting independent cycles, we get

$$|\mathrm{Vert}(S)| - 1 - (|\mathrm{Ends}(T)| - 1) - N_4 = g(T)$$
$$\implies |\mathrm{Vert}(S)| - N_4 - 1 = |\mathrm{Ends}(T)| + g(T) - 1 \;,$$

as claimed.

The geometric genus of the curves $C \in \mathrm{Sev}_n^{irr}(\Delta)$ is $g = \zeta + 1 - |\partial\Delta \cap \mathbb{Z}^2|$. Inequality (2.11) and Lemma 2.47 mean, in fact, that if a tropical curve T with Newton polygon Δ has genus $g(T) < g$, or has genus $g(T) = g$, but is not nodal, then its rank is less than ζ. Hence such curves cannot pass through Δ-generic configuration $x_1, \ldots, x_\zeta \in \mathbb{Q}^2$. Assume that T is simple, but has genus $g(T) > g$. Consider a cycle of the parameterizing trivalent graph. It corresponds to a cycle in T, formed by segments in T, joined at some trivalent vertices of T (see, for example, Figure 2.16(a), where the cycle is shown by fat lines). The cycle is dual to a fragment of the subdivision S, consisting of triangles with common edges or respectively joined by parallelograms (see Figure 2.16(b)). The limit curves, corresponding to the triangles and parallelograms in the fragment, must have components which cross the toric divisors $\mathrm{Tor}(\sigma)$, where σ runs over all the common edges of the considered triangles and parallelograms. In Figure 2.16(b) these components are depicted as fat graphs, which model the curves in the torus: one takes a tubular neighborhood of the graph in an open polygon and then its double cover ramified long the boundary. So, in the example shown in Figure 2.16(b), one has a sphere with three holes in each triangle and a cylinder in a parallelogram (cf. Sections 1.1 and 1.2, Chapter 1, presenting the relation between the algebraic curves, their complex amoebas and the non-Archimedean amoebas, or Theorem 2.21, where we glue up the complex charts of polynomials). In the deformation $C^{(t)}$, $t \in (\mathbb{C}, 0)$, the limit curves glue up and the curve $C^{(t)}$, $t \neq 0$, develops a handle, and hence one obtains at least $g(T) > g$ handles — a contradiction.

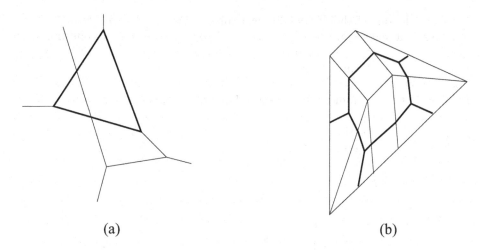

(a)   (b)

Figure 2.16: Cycles of tropical curves and handles of algebraic curves.

2.5.6 Reconstruction of a simple tropical curve

Given a Δ-general configuration of ζ points in $\mathbb{Q}^2$, there are only finitely many simple tropical curves or rank ζ with Newton polygon Δ, which pass through this configuration. Reconstruction of all such curves reduces to a linear programming problem with finite input, so can be solved by known algorithms. This approach, however, is quite costly from the computational point of view, and, moreover, does not allow one to effectively control the result.

We explain here another idea, which is due to Mikhalkin [40, 41] and which reduces the enumeration of simple tropical curves through a given configuration of points to a relatively simple combinatorial procedure, dealing with the lattice paths in the given polygon Δ (see the detailed presentation of the procedure in Section 3.6, Chapter 3).

The key ingredient is the choice of a specific configuration of fixed points. Let L be a straight line in $\mathbb{R}^2$ with the rational slope q_1/q_2, where the distinct prime numbers q_1, q_2 are much larger than the coordinates of the points in Δ. Choose generic points $p_1, p_2, \ldots, p_\zeta \in (\mathbb{K}^*)^2$ so that their valuation images in $\mathbb{R}^2$ are successive rational points $x_1, x_2, \ldots, x_\zeta$ on L satisfying

$$|x_{i+1} - x_i| \gg |x_i - x_{i-1}|, \quad \text{for all} \quad 2 \leq i < \zeta . \tag{2.14}$$

One can show that such points are in Δ-general position[5]. We want to construct a simple tropical curve T of rank ζ with Newton polygon Δ, which passes through $x_1, \ldots, x_\zeta$.

Let T be one of the desired tropical curves. Assume that besides $x_1, \ldots, x_\zeta$, it crosses L at some more points $y_1, \ldots, y_\eta$. Due to the generality conditions, the points $x_1, \ldots, x_\zeta$ must be interior points of some edges of T, and, first, we draw germs of edges of T through $x_1, \ldots, x_\zeta$. Then, at each point $y_1, \ldots, y_\eta$, we either draw a germ of an edge, or choose this point as a vertex of T and draw germs of respectively three or four adjacent edges. Further on, we orient all the edge germs (or half-germs) to be emanating from $x_1, \ldots, x_\zeta, y_1, \ldots, y_\eta$ (see Figure 2.17(a)). They are dual (and orthogonal) to some lattice segments in Δ (see Figure 2.17(b)), and hence we have only finitely many choices in the described step.

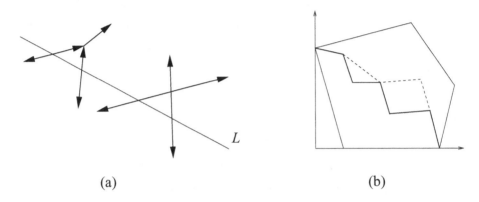

(a)	(b)

Figure 2.17: Restoring a tropical curve.

Now we extend the germs of the edges of T until appearance of an intersection point. For an intersection point we have two options (see Figure 2.17):

- either a trivalent vertex of T, and in this case the germ of the third edge is uniquely determined by the balancing condition and the choice of weights of the first two edges (notice that the latter choice is finite since the weights do not exceed the diameter of Δ),

- or a four-valent vertex of T, and then the two new edge germs of T, emanating from the vertex, are simply the continuations of the two old edges.

We naturally orient the new edges as emanating from the intersection points. Notice that the new edges always go away from L. This, in particular, implies that the extension of any newly constructed edge cannot lead to a vertex adjacent to two other edges, constructed before, and the new edges never come back to L.

---

[5]In principle, one can always slightly push the points $x_1, \ldots, x_\zeta$ from the line L to a Δ-general position, which will not affect the procedure, exposed in this section.

So, in finitely many steps we must end up. Observe that the above procedure applied to an arbitrary small variation of the configuration $x_1, \ldots, x_\zeta, y_1, \ldots, y_\eta$, will lead to a tropical curve of the same combinatorial type. Hence $\eta = 0$, since $\mathrm{rank}(T) = \zeta$. This yields that the intervals $L \backslash \{x_1, \ldots, x_\zeta\}$ are contained in the complement to the tropical curve, and thus are dual to some integral points in Δ, naturally ordered by the linear functional $\lambda(u) = \langle a, u \rangle$, where a is a directing vector of L. In turn the segments dual to the initial edge germs of the tropical curve form a λ-increasing lattice path in Δ, consisting of ζ arcs.

Then we notice that the appearance of a triple vertex of T in the above procedure is dual to the construction of a triangle from the two given adjacent edges, and the appearance of a four-valent vertex of T is dual to the construction of a parallelogram from a pair of two adjacent edges. At last, condition (2.14) assures that, on each stage, the reconstruction of a triangle or a parallelogram should be performed with a pair of edges, which is minimal with respect to the λ-ordering defined above.

Thus, we finally conclude that the described procedure of reconstruction of the required simple tropical curves is dual to the compressing procedure, which starts with a suitable lattice path in Δ consisting of ζ arcs.

2.5.7 Reconstruction of the limit curve $C^{(0)}$

The tropical curve T, constructed above, determines the function $\nu : \Delta \to \mathbb{R}$ up to an additive constant. Plugging then the coordinates of $p_m = (\xi_m(t), \eta_m(t))$ into $f(z, w) = 0$, and equating the coefficient of the minimal power of t in $f(\xi_m(t), \eta_m(t))$ to zero, we obtain

$$\sum_{(k,j) \in \delta_m} a_{kj}^{(0)} \left(\xi_m^{(0)} \right)^k \left(\eta_m^{(0)} \right)^j = 0 \, . \tag{2.15}$$

Here σ_m is the m-th edge of the lattice path and $\xi_m^{(0)}, \eta_m^{(0)}$ are the coefficients of the minimal powers of t in $\xi_m(t), \eta_m(t)$ respectively. Since the left-hand side of the above equation is a power of an irreducible binomial multiplied by a monomial, the coefficients $a_{kj}^{(0)}$, $(k, j) \in \sigma_m$, are defined uniquely up to proportionality.

Using the compressing procedure, we restore the polynomials $f_1, \ldots, f_N$ by means of the following statement.

Lemma 2.48. (i) *If Δ_m is a triangle and σ_1, σ_2 edges of Δ_m, then there are precisely $\mathrm{Area}(\Delta_m) \left(|\sigma_1| \cdot |\sigma_2| \right)^{-1}$ polynomials $f_m \in \mathcal{P}(\Delta_m)$, whose truncations $f_m^{\sigma_1}, f_m^{\sigma_2}$ are fixed and which define an irreducible rational curve in $\mathrm{Tor}(\Delta_m)$, crossing $\mathrm{Tor}(\partial \Delta_m)$ at precisely three points, where it is unibranch[6]. Moreover, all these curves are nodal and nonsingular along $\mathrm{Tor}(\partial \Delta_m)$.*

---

[6]Here $\mathrm{Area}(*)$ denotes the area normalized by the condition that the minimal lattice triangle has area 1.

(ii) *Under the hypotheses of* (i), *if* $\mathrm{Area}(\Delta_m)$ *is odd, and the truncations* $f_m^{\sigma_1}, f_m^{\sigma_2}$ *are real, then there exists precisely one real rational curve in* $\mathrm{Tor}(\Delta_m)$, *satisfying the above conditions, and all its real nodes are isolated.*

(iii) *If* Δ_m *is a parallelogram and* σ_1, σ_2 *are spanning edges of* Δ_m, *then there is a unique polynomial* $f_m \in \mathcal{P}(\Delta_m)$, *whose truncations* $f_m^{\sigma_1}$, $f_m^{\sigma_2}$ *are fixed and which splits into a product of a monomial and binomials.*

(iv) *If, under the hypotheses of* (iii), *the truncations* $f_m^{\sigma_1}$, $f_m^{\sigma_2}$ *are real, then, in the deformation* $C^{(t)}$, $t \in (\mathbb{C}, 0)$, *the nodes of* $C^{(t)}$, $t \neq 0$, *born from the intersection point of the components of the limit curve* C_m *in* $(\mathbb{C}^*)^2 \subset \mathrm{Tor}(\Delta_m)$, *all are either imaginary, or non-isolated real.*

For the proof we refer to exercises.

We say that edges σ_1, σ_2 of the subdivision S of Δ, which is dual to T, are *equivalent*, if they are opposite sides of some parallelogram Δ_k. Extend this relation by transitivity.

Thus we obtain for a given tropical curve T that the limit curve $C^{(0)}$ can be reconstructed in

$$\frac{\displaystyle\prod_{|\mathrm{Vert}(\Delta_k)|=3} \mathrm{Area}(\Delta_k)}{\displaystyle\prod_{[\sigma]} |\sigma| \cdot \prod_{k=1}^{\zeta} |\sigma_k|} \tag{2.16}$$

ways, where $[\sigma]$ runs over all aforementioned equivalence classes of edges of S.

Respectively, under the assumption that all the edges of the subdivision S have odd lattice length, we obtain a unique real tropical limit, and the contribution of the nodes, appearing from the limit curves, to the Welschinger weight is equal to

$$(-1)^s, \quad s = \#\left(\bigcup_{\substack{0 \leq k \leq N \\ |\mathrm{Vert}(\Delta_k)|=3}} (\mathrm{Int}(\Delta_k) \cap \mathbb{Z}^2) \right). \tag{2.17}$$

2.5.8 Refinement of a tropical limit

Besides the singularities of the curve $C^{(0)} = \bigcup_k C_k$ in the tori $(\mathbb{C}^*)^2 \subset \mathrm{Tor}(\Delta_k)$, $k = 1, \ldots, N$, there is another source for nodes of the curves $C^{(t)}$, $t \neq 0$, namely, the singularities of the curve $C^{(0)}$ at the toric divisors $\mathrm{Tor}(\sigma)$ for all common edges $\sigma = \Delta_k \cap \Delta_l$, $k \neq l$. We shall demonstrate this by means of the following **refinement** of the tropical limit.

Let σ be a common edge of triangles Δ_k, Δ_l from the subdivision S of Δ. If the lattice length of σ is $m \geq 2$, then by Lemma 2.46 the curves C_k, C_l are tangent to $\mathrm{Tor}(\sigma)$ at some point $p_\sigma \in \mathrm{Tor}(\sigma)$ with multiplicity m.

We perform a monomial coordinate transformation, which geometrically places the edge σ on the horizontal coordinate axis, and then (assuming that the

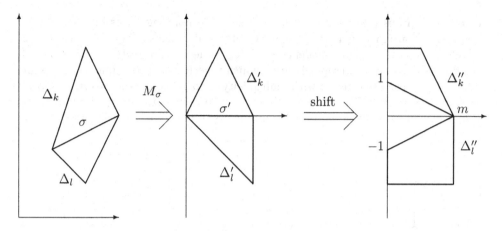

Figure 2.18: Refinement of the tropical limit at an isolated singularity.

transformed convex piecewise-linear function $\nu' : \Delta' \to \mathbb{R}$ vanishes on σ) make shift $x \mapsto x - \xi$, where ξ is the abscissa of the point p_σ. After these transformations, the polynomial $f \in \mathbb{K}[x, y]$ turns into a (Laurent) polynomial $f''(x, y)$ with some Newton polygon Δ''. The Newton polygons of the curves C_k, C_l will get a new shape as shown in Figure 2.18, leaving the triangle $\Delta_{[\sigma]} = \mathrm{conv}\{(0,1), (0,-1), (m,0)\}$ empty. The respectively changed convex piecewise-linear function ν'' is linear on Δ_k'' and Δ_l'', and in general determines some subdivision of $\Delta_{[\sigma]}$. Observe that in the proof of Lemma 2.46 we obtained that inequality (2.12) must be an equality, which, in particular, means that the singular point p_σ of $C^{(0)}$ topologically is replaced by a cylinder somehow mapped to a neighborhood of p_σ. One can show that this is possible only when ν'' is linear on $\Delta_{[\sigma]}$, and the corresponding limit curve $C_{[\sigma]}$ of f'' with Newton triangle $\Delta_{[\sigma]}$ is rational. Furthermore, we claim

Lemma 2.49. *In the above notation, given non-zero coefficients at the vertices of $\Delta_{[\sigma]}$, there are precisely m (Laurent) polynomials in $\mathcal{P}(\Delta_{[\sigma]})$, which have zero coefficients at $(m - 1, 0)$ and define rational curve $C_{[\sigma]}$ in $\mathrm{Tor}(\Delta_{[\sigma]})$. Each of these curves has $m - 1$ nodes as only singularities.*

Furthermore, if the given coefficients are real, then

(i) *for even m, either there is no real rational curve $C_{[\sigma]}$ as above, or there are precisely two such real curves: one of them has $m - 1$ isolated real nodes, the other has one non-isolated real node and $m - 2$ imaginary nodes;*

(ii) *for odd m, there is precisely one real rational curve $C_{[\sigma]}$ as above, and it has $m - 1$ isolated real nodes.*

The proof is left to the reader as an exercise.

A little bit more complicated treatment is required for the case of a non-isolated singularity of $C^{(0)}$. In this situation we have triangles Δ_k, Δ_l joined by a sequence of parallelograms (see Figure 2.19(a)). The corresponding edges σ, σ' of Δ_k, Δ_l, respectively, are included into a sequence of parallel edges of equal length $m \geq 2$, which all constitute an equivalence class $[\sigma]$ as defined in Section 2.5.7 (they all are dual to edges of the tropical curve, which together form a segment). The limit curves C_k and C_l are tangent with multiplicity m to $\mathrm{Tor}(\sigma)$, respectively to $\mathrm{Tor}(\sigma')$ at some points $p_\sigma \in \mathrm{Tor}(\sigma)$, $p_{\sigma'} \in \mathrm{Tor}(\sigma')$. The points p_σ and $p_{\sigma'}$ are joined by a sequence of m-multiple binomial components of the limit curves, corresponding to the parallelograms.

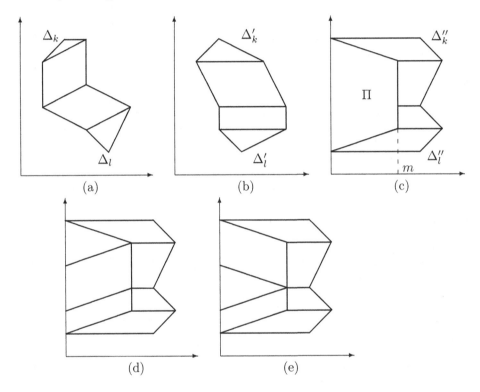

Figure 2.19: Refinement of the tropical limit along a non-isolated singularity.

Again we perform a monomial transformation making σ and σ' horizontal (Figure 2.19(b)). The intersection points of $C^{(0)}$ with the toric divisors, corresponding to the edges in the equivalence class $[\sigma]$, have the same abscissa ξ, and thus, performing the shift $x \mapsto x - \xi$ (and assuming without loss of generality that the convex function ν is constant along σ), we obtain new shapes of the Newton polygons of the limit curves, which leave empty the trapeze Π (see Figure 2.19(c)). The polynomial $f(x, y) \in \mathbb{K}[x, y]$, defining the curve C, transforms into a polynomial $f''(x, y)$ with Newton polygon Δ''. The tropical limit of f'' gives a convex piecewise-linear function $\nu'' : \Delta'' \to \mathbb{R}$, defining a subdivision of Π, and determines new limit curves with Newton polygons inside Π. As above, the equality in (2.12), derived in the proof of Lemma 2.46, requires that the components of the new limit curves, which cross the toric divisors corresponding to segments on the right vertical side of Π, must be binomial, and in the deformation $C^{(t)}$, $t \in (\mathbb{C}, 0)$, they do not glue up with any other components of these limit curves. Furthermore, the other components must be rational too. One can show that then the only possible subdivision of Π, satisfying these requirements, consists of parallelograms and one triangle $\Delta_{[\sigma]}$ with the vertical side of length 2 (see, for example, Figure 2.19(d,e)), in turn, the limit curves in parallelograms consist of only binomial components, whereas the limit curve $C_{[\sigma]}$ with Newton triangle $\Delta_{[\sigma]}$ is as described in Lemma 2.49. Notice that the function ν'' is linear on each of the Newton polygons surrounding Π (see Figure 2.19(c)), and it admits a unique extension as a convex piecewise-linear function determining a subdivision of Π into parallelograms and one triangle.

Remark 2.50. The refinement procedure described here can be viewed as follows. In case of an isolated singular point $p_\sigma \in \mathrm{Tor}(\sigma)$ of $C^{(0)}$, we make a weighted blow-up of the point p_σ of the threefold $\mathrm{Tor}(\widetilde{\Delta})$, so that there appears an exceptional divisor $\mathrm{Tor}(\Delta_{[\sigma]})$. In case of a non-isolated singularity of $C^{(0)}$ we, in fact, make a combination of blow-ups along the multiple binomial components so that there appears a sequence of exceptional toric divisors associated with the polygons, forming a subdivision of Π.

For any equivalence class $[\sigma]$ of the edges of the subdivision S (as defined in Section 2.5.7) a pair $(\Delta_{[\sigma]}, C_{[\sigma]})$ is called a **refinement of the tropical limit of f at** $[\sigma]$. Then we construct the **refined tropical limit** of a curve $C \in |\mathcal{L}_\Delta|$ by appending the triangles $\Delta_{[\sigma]}$ to the combinatorial part of the tropical limit and by appending the limit curves $C_{[\sigma]}$ to the sequence $C_1, \ldots, C_N$.

In view of (2.16) and Lemma 2.49, the number of the refined tropical limits associated with the curve C is equal to

$$\frac{\prod_{|\mathrm{Vert}(\Delta_k)|=3} \mathrm{Area}(\Delta_k)}{\prod_{k=1}^r |\sigma_k|} . \tag{2.18}$$

If C is defined over $\mathbb{K}_\mathbb{R} \overset{\mathrm{def}}{=} \bigcup_{m \geq 1} \mathbb{R}\{\{t^{1/m}\}\} \subset \mathbb{K}$ then, taking into account (2.17 and Lemma 2.49, we decide that the total contribution of the possible refined tropical limits of C into the Welschinger invariant is zero as far as there is at least one even length edge in the subdivision S, or is given by (2.17), if all the edges of S have odd length.

2.5.9 Refinement of the condition to pass through a fixed point

We have used the fixed point condition in Section 2.5.7, taking into account the "zero order approximation" (2.15). Now we refine the condition to pass through a fixed point by using the next approximation.

Assume that an edge σ_m of the lattice path has length $d \geq 2$. After a suitable monomial transformation, we can assume that the edge σ_m becomes $[(0,0),(d,0)]$ and $\nu\big|_{\sigma_m} = 0$. That is

$$f(z,w) = \sum_{k=0}^{d} \left(a_{k,0}^{(0)} + c_{k,0}(t) \right) z^k + O(t) \ ,$$

where $c_{k,0}(0) = 0$. Furthermore,

$$f(z,w) = \sum_{k=0}^{d} c_{k,0}(t) z^k + a_{k,0}^{(0)}(z - \xi_m^{(0)})^d + O(t) \ .$$

Letting $z' = z - \xi_m^{(0)}$, we obtain

$$f'(z',w) = \sum_{k=0}^{d} c_{k,0}'(t) z'^k + \left(a_{k,0}^{(0)} + c_{d,0}(t) \right) (z')^d + O(t)$$

with $c_{k,0}'(0) = 0$. Making the variable change $z' = z'' + \tau$, where

$$\tau = -\frac{c_{d-1,0}'(t)}{d \cdot a_{d,0}^{(0)}} + \text{h.o.t.}$$

is chosen so that the polynomial

$$f''(z'',w) \overset{\text{def}}{=} f'(z',w)$$

has coefficient 0 at $(z'')^{d-1}$, we end up with

$$f''(z'',w) = \sum_{k=0}^{d-2} c_{k,0}''(t) \cdot z''^k + (a_{d,0}^{(0)} + c_{d,0}''(t))(z'')^k$$
$$+ w \cdot \left(a_{0,1}^{(0)} + c_{0,1}(t) \right) t^r + w^{-1} \cdot \left(a_{0,-1}^{(0)} + c_{0,-1}(t) \right) t^s + \text{h.o.t.} \ ,$$

where $c_{k,j}''(0) = 0$, and $r \neq s$ (say, $r < s$). Then we plug the new coordinates

$$(t\xi_m^1 - \tau + \cdots, \quad \eta_m^{(0)} + \eta_m^1 \cdot t + \cdots)$$

of p_m into the equation $f''(z'',w) = 0$, and equate to zero the minimal power of t, which appears just in the coefficient of $(z'')^d$ and y. This gives

$$\eta_m^{(0)} a_{0,1}^{(0)} t^r + a_{d,0}^{(0)}(\xi_m^1 t - \tau)^d + \text{h.o.t.} = 0 \ ,$$

and hence

$$\tau = \xi_m^1 t - \left(-\frac{\eta_m^{(0)} a_{0,1}^{(0)}}{a_{d,0}^{(0)}} \right)^{\frac{1}{d}} t^{\frac{r}{d}} + \text{h.o.t..} \tag{2.19}$$

Finally we get

$$c_{0,0}(t) - \frac{a_{0,0}^{(0)}}{a_{d,0}^{(0)}} c_{d,0}(t) = (-1)^d d a_{d,0}^{(0)} (\xi_m^{(0)})^{d-1} \xi_m^1 t$$

$$+ (-1)^{d-1} d (\xi_m^{(0)})^{d-1} \left(-\eta_m^{(0)} a_{0,1}^{(0)} (a_{d,0}^{(0)})^{d-1} \right)^{\frac{1}{d}} t^{\frac{r}{d}} + \text{h.o.t.} \qquad (2.20)$$

which gives d options for the left-hand side. In the real case, if d is even, we obtain either no real refinements or two real refinements, if d is odd, we always have one real refinement.

Combining the above conclusion with (2.17) and (2.18), we get, for a given tropical curve T, precisely

$$\prod_{|\Delta_k|=3} (\text{Area}(\Delta_k))$$

choices of the refined topical limits with refined conditions to pass through the given points $p_1, p_2, \ldots, p_\zeta$.

2.5.10 Refined patchworking theorem

The last step in the proof of Mikhalkin's correspondence theorem and in the tropical formula for the Welschinger invariants is the following

Theorem 2.51 (Refined patchworking theorem). *Let a generic configuration $p_1, \ldots, p_\zeta$ in $(\mathbb{K}^*)^2$ project by valuation to a Δ-general configuration $x_1, \ldots, x_\zeta \in \mathbb{Q}^2$. Let us be given*

- *a simple irreducible tropical curve T of rank ζ with Newton polygon Δ, passing through $x_1, \ldots, x_\zeta \in \mathbb{Q}^2$, and dual to a subdivision $S : \Delta = \Delta_1 \cup \cdots \cup \Delta_N$;*

- *curves $C_m \in |\mathcal{L}_{\Delta_m}|$, $m = 1, \ldots, N$, which satisfy the conditions of Lemma 2.48, are compatible to each other in the sense that $C_m \cap \text{Tor}(\sigma) = C_l \cap \text{Tor}(\sigma)$ as $\sigma = \Delta_m \cap \Delta_l$, and are compatible with the points $p_1, \ldots, p_\zeta$ by (2.15);*

- *curves $C_{[\sigma]}$, compatible with $C_1, \ldots, C_N$, satisfying the conditions of Lemma 2.49, and associated with all equivalence classes of edges of S;*

- *refined conditions to pass through the points $p_1, \ldots, p_\zeta$ described by (2.20) with specified roots of the coefficients of powers of t in the right-hand side.*

Then there exists a unique irreducible curve $C \in |\mathcal{L}_\Delta|_{\mathbb{K}}$ with n nodes as its only singularities, passing through $p_1, \ldots, p_\zeta$, and such that its refined tropical limit and refined conditions to pass through the fixed points fit the given data.

Furthermore, if the given data are real, then the curve C is defined over $\mathbb{K}_{\mathbb{R}}$.

We do not set forth a detailed proof, but discuss the general strategy and comment on important points.

We define the required curve C by a polynomial

$$f(x,y) = \sum_{(k,j)\in\Delta} a_{kj}(t)x^k y^j t^{\nu(i,j)}, \quad a_{kj}(t) = a_{kj}^{(0)} + c_{kj}(t),\ c_{kj}(0) = 0,\ (k,j)\in\Delta,$$
(2.21)

where

$$C_m = \left\{ f_m(x,y) \stackrel{\text{def}}{=} \sum_{(k,j)\in\Delta_m} a_{kj}^{(0)} x^k y^j = 0 \right\}, \quad m = 1,\dots,N.$$

For any $m = 1,\dots,N$, take the linear function $\lambda_m : \Delta \to \mathbb{R}$ equal to ν on Δ_m, and put $\nu_m = \nu - \lambda_m$. Similarly, for any edge σ of S, take a linear function $\lambda_\sigma : \Delta \to \mathbb{R}$ equal to ν along σ and strongly less than ν outside σ; put $\nu_\sigma = \nu - \lambda_\sigma$. Substitution of ν_k or ν_σ for ν in (2.21) is equivalent to a coordinate change (cf. proof of Theorem 2.35).

For a given configuration $x_1,\dots,x_\zeta$, a given combinatorial type of the tropical curve T and a given combinatorial type of the distribution of the points $x_1,\dots,x_\zeta$ on T, we perform the reconstruction procedure for T similar to that from Section 2.5.6. Namely, we start with germs of edges of T passing through $x_1,\dots,x_\zeta$. Denote by $\sigma_1,\dots,\sigma_\zeta$ the dual edges of the subdivision S, respectively. Then we extend these germs until we come to the first possible intersection point. It is a vertex of T, whose dual polygon of S we denote by Δ_1. Since we know whether it is a triangle or a parallelogram, we can uniquely restore T in a neighborhood of that vertex. Then we continue the extension of edges until we come to the next possible intersection point, dual respectively to Δ_2, and so on. The tropical curve T is determined by the configuration $x_1,\dots,x_\zeta$ and the given combinatorial types, and thus, the procedure completely restores T. Considering this procedure as a reconstruction of the subdivision S, we observe that each polygon Δ_m is built on two of its sides, which have appeared before. We denote the union of these sides by $(\partial\Delta_m)_+$. We can also pick up an endpoint $(\partial\sigma_k)_+$ of each edge σ_k, $1 \le k \le \zeta$ so that the sets $\sigma_k\backslash(\partial\sigma_k)_+$, $k = 1,\dots,\zeta$, will be disjoint.

Our plan is as follows: we split the linear polynomial space $\mathcal{P}(\Delta)$ generated by the monomials $x^k y^j$, $(k,j) \in \Delta$, into the direct sum of subspaces, and then show that each of them is "responsible" for the appearance of some nodes of the curves $C^{(t)}$, $t \ne 0$, which we want to construct.

1. Let Δ_m be a triangle. The polynomial $f_m(x,y)$ defines a rational curve $C_m \subset \mathrm{Tor}(\Delta_m)$ with $|\mathrm{Int}(\Delta_m) \cap \mathbb{Z}^2|$ nodes in $(\mathbb{C}^*)^2$. The polynomial

$$f_{(t),m}(x,y) = \sum_{(k,j)\in\Delta} a_{kj}(t)x^k y^j t^{\nu_m(k,j)} \in \mathcal{P}(\Delta)$$

is a small deformation of f_m in $\mathcal{P}(\Delta)$ (cf. the proof of Theorem 2.35), and we impose the condition to keep the nodes of C_m in the deformation $C^{(t)}$, $t \in (\mathbb{C}, 0)$, by means of

Lemma 2.52. *The germ at f_m of the set of polynomials in $\mathcal{P}(\Delta)$, defining a curve with $|\operatorname{Int}(\Delta_m) \cap \mathbb{Z}^2|$ nodes in a neighborhood of the nodes of C_m in $(\mathbb{C}^*)^2$, is smooth, has codimension $|\operatorname{Int}(\Delta_m) \cap \mathbb{Z}^2|$, and intersects transversally with the space $\mathcal{P}(\Delta_m, \partial \Delta_m, f_m)$ at one point (i.e., at $\{f_m\}$).*

This statement is a sort of the transversality conditions discussed in Sections 2.4.2 and 2.4.4, and we leave the proof as exercise.

2. Let Δ_m be a parallelogram. The curve C_m contains two distinct binomial components C_{m1}, C_{m2} of multiplicities d_1, d_2, respectively. Notice that the total intersection multiplicity of $C_{m1}^{d_1}$ and $C_{m2}^{d_2}$ in $(\mathbb{C}^*)^2$ is equal to $|(\Delta_m \backslash (\partial \Delta_m)_+) \cap \mathbb{Z}^2|$. As in the proof of Lemma 2.46, in the deformation $C^{(t)}$, $t \in (\mathbb{C}, 0)$, all these intersection points must turn in $|(\Delta_m \backslash (\partial \Delta_m)_+) \cap \mathbb{Z}^2|$ nodes of $C^{(t)}$, $t \neq 0$, in $(\mathbb{C}^*)^2$. We treat this condition by means of

Lemma 2.53. *The germ at f_m of the set of polynomials in $\mathcal{P}(\Delta)$, defining a curve with $|(\Delta_m \backslash (\partial \Delta_m)_+) \cap \mathbb{Z}^2|$ nodes in a neighborhood of the intersection points of the components C_{m1} and C_{m2} in $(\mathbb{C}^*)^2$ is smooth, has codimension $|(\Delta_m \backslash (\partial \Delta_m)_+) \cap \mathbb{Z}^2|$, and intersects transversally with the space $\mathcal{P}(\Delta_m, (\partial \Delta_m)_+, f_m)$ at one point (i.e., at $\{f_m\}$).*

Again the proof is left as an exercise.

3. Consider the edge $\sigma = \sigma_m$ dual to the edge of T, passing through the point x_m, $1 \leq m \leq \zeta$. Performing if necessary a monomial transformation like that in Section 2.5.8, we can assume that

$$f_\sigma(x,y) = \sum_{(k,j) \in \sigma} a_{kj}^{(0)} x^k y^j = (x + \xi)^d, \quad d = |\sigma|, \tag{2.22}$$

and our requirement is that the polynomial

$$f_{(t),\sigma}(x,y) = \sum_{(k,j) \in \Delta} a_{kj}(t) x^k y^j t^{\nu_\sigma(k,j)}, \tag{2.23}$$

which is a small deformation of f_σ, defines curves $C^{(t)}$, $t \neq 0$, having $d - 1$ nodes in a neighborhood of the point $(-\xi, 0)$, and satisfies the condition $f_{(t),\sigma}(p_m) = 0$. Specifying refinements of the tropical limit of $f_{(t),\sigma}$, we shall express these requirements in the form of relations on the coefficients of the considered polynomials. Denote by U the space of polynomials in $\mathbb{K}[x,y]$ given by the right-hand side of (2.23), where

$$a_{kj}(t) = a_{kj}^{(0)} + c_{kj}(t), \quad c_{kj}(0) = 0, \quad (k,j) \in \Delta \cap \mathbb{Z}^2. \tag{2.24}$$

Lemma 2.54. *In the above notation, let $V_{[\sigma]}$ be the set of polynomials $F \in U$ such that*

- *the refinement of the tropical limit of F at $[\sigma]$ is a given pair $(\Delta_{[\sigma]}, C_{[\sigma]})$,*

- *F defines a family of curves $C^{(t)}$, $t \in (\mathbb{C}, 0)$, in $\mathbb{C}^2 \backslash \{x = 0\}$ having $d - 1$ nodes in a neighborhood of the point $(\xi, 0)$,*

- *F satisfies the condition $F(p_m) = 0$ refined to (2.20) with a fixed d-th root of the expression $(-\xi_m^{(0)} a_{0,1}^{(0)} (a_{d,0}^{(0)})^{d-1})$ in the right-hand side.*

Then $V_{[\sigma]}$ is given by relations

$$c_{kj}(t) = \Psi_{kj}\left(\{c_{k'j'}(t) \ : \ (k', j') \in \Delta \backslash (\sigma \backslash \partial\sigma_+)\}, \ t\right), \quad (k, j) \in \sigma \backslash \partial\sigma_+ \ ,$$

where Ψ_{kj} are some analytic functions in a neighborhood of the origin such that $\Psi_{kj}(0) = 0$, $(k, j) \in \sigma \backslash \partial\sigma_+$.

We do not prove this statement, but should like to comment on it, since it differs from the previous Lemmas 2.52 and 2.53. After the shift

$$x \mapsto x - \xi - \tau \tag{2.25}$$

with some $\tau = \tau(t)$, $\tau(0) = 0$, the polynomial $F(x, y)$ given by (2.24) turns into a new polynomial $F''(x, y)$, which we consider over $\mathbb{K}$, and whose tropical limit contains a convex piecewise-linear function ν'' with the triangle $\Delta_{[\sigma]} = \mathrm{conv}\{(0, 1), (0, -1), (d, 0)\}$ as one of the linearity domains (cf. Section 2.5.8). Notice that the function ν'' is uniquely determined by the initial data of Theorem 2.51. Furthermore, assuming that ν'' vanishes on $\Delta_{[\sigma]}$, we see that

$$F''(x, y) = \sum_{(k,j) \in \Delta_{[\sigma]}} (c_{kj}^{(0)} + c_{kj}(t)) x^k y^j + O(t), \quad c_{ij}(0) = 0, \ (i, j) \in \Delta_{[\sigma]} \ ,$$

where $O(t)$ contains monomials from outside of $\Delta_{[\sigma]}$, and

$$\Phi(x, y) = \sum_{(k,j) \in \Delta_{[\sigma]}} b_{kj}^{(0)} x^k y^j$$

is an equation of $C_{[\sigma]}$, in which we assume $b_{d-1,0}^{(0)} = 0$. Then the problem reduces to a transversality statement like in Lemmas 2.52 and 2.53 with an additional equation (2.19) for $\tau(t)$ (in which one has to specify the d-th root).

4. Let σ be en edge of S, which is not equivalent to any of $\sigma_1, \ldots, \sigma_\zeta$, and has length $d \geq 2$. The equivalence class $[\sigma]$ is totally ordered by the reconstruction procedure for T, and we suppose that σ is the first in its class. As above we can assume that σ lies on the horizontal axis, and that $f_\sigma = \sum_{(k,j) \in \sigma} a_{kj}^{(0)} x^k y^j$ satisfies (2.22). We require that the polynomial $f_{(t),\sigma}(x, y)$, given by (2.23), defines curves $C^{(t)} \subset \mathrm{Tor}(\Delta)$, $t \neq 0$, having $d - 1$ nodes in a neighborhood of $(-\xi, 0)$. Similar to the preceding step, we specify refinement of the tropical limit of $f_{(t),\sigma}$ and express the above requirement in the following analytic form.

Lemma 2.55. *In the above notation, let $V_{[\sigma]}$ be the set of polynomials $F \in U$ such that*

- *the refinement of the tropical limit of F at $[\sigma]$ is a given pair $(\Delta_{[\sigma]}, C_{[\sigma]})$,*

- *F defines a family of curves $C^{(t)}$, $t \in (\mathbb{C}, 0)$, in $\mathbb{C}^2 \setminus \{x = 0\}$ having $d - 1$ nodes in a neighborhood of the point $(\xi, 0)$.*

Then $V_{[\sigma]}$ is given by relations

$$c_{kj}(t) = \Psi_{kj}\left(\{c_{k'j'}(t) \ : \ (k', j') \in \Delta \setminus (\sigma \setminus \partial\sigma)\}, \ t\right), \quad (k, j) \in \sigma \setminus \partial\sigma ,$$

where Ψ_{kj} are some analytic functions in a neighborhood of the origin such that $\Psi_{kj}(0) = 0$, $(k, j) \in \sigma \setminus \partial\sigma$.

The statement is similar to Lemma 2.54. We only remark, that the edge σ appears in the reconstruction procedure for S (dual to that for T) as the third side of some triangle Δ_s, and then the parameter τ from (2.25) is, in fact, determined by the transversality conditions for Δ_s from Lemma 2.52.

5. Observe that the set of integral points of Δ is split into disjoint subsets

- $(\Delta_m \setminus \partial\Delta_m) \cap \mathbb{Z}^2$ for all triangles Δ_m,

- $(\Delta_m \setminus (\partial\Delta_m)_+) \cap \mathbb{Z}^2$ for all parallelograms Δ_m,

- $(\sigma_m \setminus (\partial\sigma_m)_+) \cap \mathbb{Z}^2$, $m = 1, \ldots, \zeta$,

- $(\sigma \setminus \partial\sigma) \cap \mathbb{Z}^2$ for the first edge σ in each equivalence class $[\sigma]$ disjoint with $\{\sigma_1, \ldots, \sigma_\zeta\}$.

Correspondingly the hyperplane in $\mathcal{P}(\Delta)$ defined by the vanishing coefficient at the endpoint $(\partial\sigma_1)_+$ of σ_1, splits into the direct sum of linear subspaces, which are ordered by the reconstruction procedure for T. In the same manner as in the proof of Theorem 2.35 (see Remark 2.36), the conditions of Lemmas 2.52, 2.53, 2.54, and 2.55 lead to a system of equations for the coefficients $c_{kj}(t)$, $(k, j) \in \Delta \setminus (\partial\sigma_1)_+$, with a block-triangular linearization, and then we decide on the existence and uniqueness of a solution by the implicit function theorem.

2.5.11   The real case: Welschinger invariants

Now we explain how to obtain the tropical formula for the Welschinger invariants.

Again we state the problem over the field $\mathbb{K}$, possessing the natural complex conjugation and containing the real subfield $\mathbb{K}_{\mathbb{R}} \subset \mathbb{K}$.

Under the hypotheses of Theorem 2.44, we derive from Lemma 2.46 that the tropical limit of a real rational curve in $\text{Tor}_{\mathbb{K}}(\Delta)$, passing through generic $\zeta = |\partial\Delta| - 1$ real points, consists of an irreducible simple tropical curve with Newton polygon Δ of rank ζ, and a collection of real curves as specified in Lemma 2.46(2).

Next we notice that by Theorem 2.51, the data consisting of (i) an irreducible simple tropical curve T with Newton polygon Δ of rank ζ, (ii) an appropriate collection of real curves as in the assertion of Theorem 2.51, and (iii) the real refined conditions to pass through the fixed points (2.20), produce a unique rational curve $C \in |\mathcal{L}_\Delta|$, defined over $\mathbb{K}_\mathbb{R}$ and passing through the fixed points. Furthermore, the nodes of C come

- from the nodes of the real rational curves C_m such that Δ_m is a triangle in the subdivision S of Δ,

- from the nodes of the curves $C_{[\sigma]}$ with $[\sigma]$ running over all the equivalence classes of edges of S,

- from the intersection points of distinct binomial components of the curves C_m such that Δ_m is a parallelogram in S; in this case no real solitary node may appear.

Observe that the choice of the curves $C_{[\sigma]}$, specified in Theorem 2.51, is independent of the refined conditions to pass through the fixed points, provided that the curves $C_m \in |\mathcal{L}_{\Delta_m}|$, $m = 1, \ldots, N$, are already given. Hence, if there is an edge σ of S of an even length d, then by Lemma 2.49, either there exist no suitable real curves $C_{[\sigma]}$, or there are two suitable real curves $C_{[\sigma]}$, having distinct parity of the number of real solitary nodes. Thus, as noticed above, the real rational curves C, which can be constructed out of a given tropical curve T, in total contribute 0 to the Welschinger invariant, which agrees with the definition $\mathcal{W}(T) = 0$ in this case.

If all the edges of S have odd length, then by Lemmas 2.48 and 2.49, and by formula (2.20), there is a unique choice of appropriate real curves C_m, $m = 1, \ldots, N$, and $C_{[\sigma]}$, $\sigma \in \mathrm{Edges}(S)$, and the refined conditions to pass through the fixed point. Hence there exists a unique real rational curve $C \in |\mathcal{L}_\Delta|$, passing through $p_1, \ldots, p_\zeta$ and projecting to T. By Lemmas 2.48 and 2.49 the total parity of the number of real solitary nodes of the curves C_m and $C_{[\sigma]}$ coincides with the parity of the number of interior integral points in all the triangles of the subdivision S, which means that the contribution to the Welschinger invariant of the tropical curve T is equal to $\mathcal{W}(T) = (-1)^{s(T)}$.

2.6   Exercises

Exercise 2.1. Given a square $ABCD$, find a convex subdivision of the triangle ABD and a convex subdivision of the triangle BCD, which together form a non-convex subdivision of the given square.

Exercise 2.2. Using the combinatorial patchworking, construct for any positive integer k

- a curve C_0 of degree $2k$ in $\mathbb{R}P^2$ such that the real point set of C_0 is empty,

- a curve C_1 of degree $2k - 1$ in $\mathbb{R}P^2$ such that the real point set of C_1 is connected.

Exercise 2.3. Let m be a positive integer, and T the triangle in $\mathbb{R}^2$ with the vertices $(0, 0)$, $(0, m)$, and $(m, 0)$. Find a primitive triangulation of T and a sign distribution at the vertices of this triangulation which produce, via the combinatorial patchworking, a curve of degree m in $\mathbb{R}P^2$ such that the real point set of this curve is connected.

Exercise 2.4. Let m be a positive integer, and T the triangle in $\mathbb{R}^2$ with the vertices $(0, 0)$, $(0, m)$, and $(m, 0)$. Assume that any integral point of T is endowed with the sign $+$.

- Find a primitive convex triangulation of T which together with the given sign distribution produces, via the combinatorial patchworking, a hyperbolic curve of degree m in $\mathbb{R}P^2$ (a real curve of degree m in $\mathbb{R}P^2$ is called *hyperbolic* if there is a real point such that any real line through this point crosses the curve only at real points).

- Find a primitive convex triangulation of T which together with the given sign distribution produces, via the combinatorial patchworking, an M-curve of degree m in $\mathbb{R}P^2$.

Exercise 2.5. Construct Harnack's, Hilbert's and Gudkov's M-curves of degree 6 in $\mathbb{R}P^2$ using the patchworking method and charts of real cubics.

Exercise 2.6. Show that the lattice subdivision of the plane shown in Figure 2.20 is convex. Using this subdivision and an appropriate patchworking theorem, construct for any degree $d \geq 3$ a real algebraic plane curve of degree d having $[(d^2 - 3d + 4)/4]$ real cusps as its only singularities.

Exercise 2.7. Let m be a positive integer, $\Delta \subset \mathbb{R}^3$ the tetrahedron with vertices $(0, 0, 0)$, $(m, 0, 0)$, $(0, m, 0)$, and $(0, 0, m)$, and S a T-surface of degree m in $\mathbb{R}P^3$ constructed by means of a primitive triangulation of Δ. Prove that the Euler characteristic $\chi(\mathbb{R}S)$ of the real point set $\mathbb{R}S$ of S satisfies the equality

$$\chi(\mathbb{R}S) = -\frac{m^3}{3} + \frac{4m}{3} \ .$$

(Note that $-\frac{m^3}{3} + \frac{4m}{3}$ is the signature of the complex point set $\mathbb{C}S$ of S in $\mathbb{C}P^3$.)

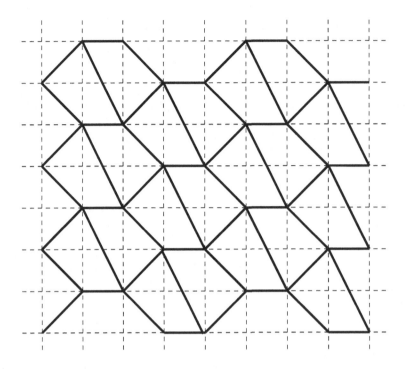

Figure 2.20: Lattice subdivision for Exercise 2.6.

Exercise 2.8. Let $A \subset \mathbb{R}_+^n$ be a set of vertices of a non-degenerate lattice n-simplex Δ. Assume that no hyperplane through the origin contains a facet of Δ. Prove that a real polynomial $f(x) = \sum_{\omega \in A} a_\omega x^\omega$ has at most one real critical point in the positive orthant $\mathbb{R}_+^n$; moreover, such a point is non-degenerate and has index i, or $n - i$, where i is the number of facets of Δ which are visible from the origin.

Chapter 3

Applications of tropical geometry to enumerative geometry

3.1 Introduction

The main purpose of this chapter is to present several applications of tropical geometry in enumerative geometry. The idea to use tropical curves in enumerative questions, and in particular in classical questions of enumeration of algebraic curves (satisfying some constraints) in algebraic varieties was suggested by M. Kontsevich. This idea was realized by G. Mikhalkin [40, 42] who established an appropriate correspondence theorem between the complex algebraic world and the tropical one. This correspondence allows one to calculate Gromov–Witten type invariants of toric surfaces, namely, to enumerate certain nodal complex curves of a given genus which pass through given points in a general position in a toric surface. Roughly speaking, Mikhalkin's theorem affirms that the number of complex curves in question is equal to the number of their tropical analogs passing through given points in a general position in $\mathbb{R}^2$ and counted with multiplicities. In addition, [40] suggests a combinatorial algorithm for an enumeration of the required tropical curves. An extension of Mikhalkin's correspondence theorem to the case of rational curves in toric varieties was proposed by T. Nishinou and B. Siebert in [48].

The tropical approach has important applications in enumerative real algebraic geometry as well. Enumerative geometric problems over the reals, such as counting real algebraic curves in real algebraic varieties, have a different character than in the complex case, since over the reals the answer typically depends on the configuration of the imposed constraints. Thus, the main question concerns the upper and lower bounds for the number of real solutions. For these problems of counting curves, the corresponding Gromov–Witten type invariant (i.e., the number of the corresponding complex curves) is an upper bound. No non-trivial lower bound was known until the recent discovery by J.-Y. Welschinger [73, 74, 75] of invariants which can be seen as real analogs of genus zero Gromov–Witten invariants. The theory of the Welschinger invariants is under an intensive development. The tropical approach allows one to calculate or estimate these invariants in some situations which leads to surprising results in enumerative real algebraic geometry (for example, the logarithmic equivalence of the Gromov–Witten and the Welschinger invariants of toric Del Pezzo surfaces [28]).

3.2   Tropical hypersurfaces in $\mathbb{R}^n$

We briefly recall here the definition and the basic properties of tropical hypersurfaces in $\mathbb{R}^n$ (in fact, $\mathbb{R}^n$ is the tropical analog of the complex torus $(\mathbb{C}^*)^n$ and should be viewed here as $(\mathbb{T}^*)^n = (\mathbb{T} \setminus \{-\infty\})^n$; cf. Chapter 1, Section 1.5).

Fix a positive integer n. A point in $\mathbb{R}^n$ is called *integer*, if all coordinates of this point are integer. Let A be a finite collection of integer points in $\mathbb{R}^n$, and $\varphi : A \to \mathbb{R}$ an arbitrary function. The pair (A, φ) gives rise to a **tropical hypersurface** in $\mathbb{R}^n$ in the following way.

Let $\widehat{\varphi} : \mathbb{R}^n \to \mathbb{R}$ be the *Legendre transform* of φ:

$$\widehat{\varphi}(x_1, \ldots, x_n) = \max_{(i_1, \ldots, i_n) \in A} \{i_1 x_1 + \cdots + i_n x_n - \varphi(i_1, \ldots, i_n)\}.$$

The function $\widehat{\varphi}$ is a *tropical polynomial* defining the hypersurface under description (cf. Chapter 1, Section 1.5). Notice that $\widehat{\varphi}$ is convex piecewise-linear, and consider the corner locus $T(A, \varphi)$ of $\widehat{\varphi}$, i.e., the subset of $\mathbb{R}^n$ formed by the points where $\widehat{\varphi}$ is not locally affine-linear. The graph $\Gamma(A, \varphi)$ of $\widehat{\varphi}$ is naturally stratified. The set $T(A, \varphi)$ is stratified by the projections of the elements of the stratification of $\Gamma(A, \varphi)$, and defines a subdivision $\Theta(A, \varphi)$ of $\mathbb{R}^n$. The $(n-1)$-dimensional elements of the stratification of $T(A, \varphi)$ are called *facets*.

Each facet σ of $T(A, \varphi)$ can be equipped with a positive integer number. Namely, let σ be the projection of an $(n-1)$-dimensional polyhedron Σ in $\Gamma(A, \varphi)$. Denote by A_σ the subset of A formed by the points $(i_1, \ldots, i_n)$ such that the graph of the affine-linear function $i_1 x_1 + \cdots + i_n x_n - \varphi(i_1, \ldots, i_n)$ contains Σ. Notice that A_σ has at least two points, and all the points of A_σ belong to a straight line. Denote by I and J the two extremal points of A_σ, i.e., the points of A_σ such that the segment $[IJ]$ contains all the points of A_σ. Associate to the facet σ a *weight* $w(\sigma)$ equal to the integer length of $[IJ]$ (the *integer length* of a segment with integer endpoints is the number of its integer points diminished by 1).

Definition 3.1. The polyhedral complex $T(A, \varphi)$ whose facets are equipped with the corresponding weights is called the *tropical hypersurface* associated with the

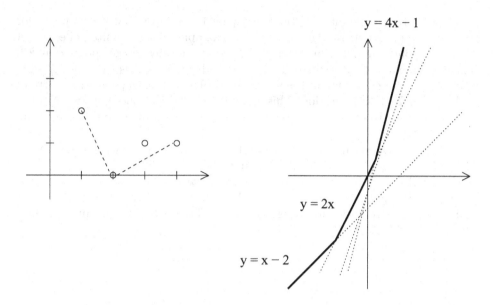

Figure 3.1: Legendre transform in dimension 1.

pair (A, φ). One says that $T(A, \varphi)$ is a tropical hypersurface with Newton polytope $\Delta(A)$, where $\Delta(A)$ is the convex hull of A. If $\Delta(A)$ is the simplex with vertices $(0, 0, \ldots, 0)$, $(m, 0, \ldots, 0)$, $(0, m, 0, \ldots, 0)$, $\ldots$, $(0, \ldots, 0, m)$, then the tropical hypersurface $T(A, f)$ is said to be *of degree m*.

Example 3.2. Let $A = \{(0, 0), (1, 0), (0, 1)\} \subset \mathbb{R}^2$, and $\varphi : A \to \mathbb{R}$ an arbitrary function. The tropical curve associated with the pair (A, φ) is the union of three rays in $\mathbb{R}^2$ which share a common extremal point; the directions of the rays are south, west and northeast (see Figure 3.2). In this case, a change of values of φ leads to a translation of the tropical curve. The common point of the three rays has the coordinates $(\varphi(1, 0) - \varphi(0, 0), \varphi(0, 1) - \varphi(0, 0))$.

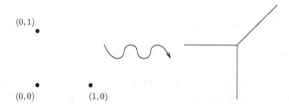

Figure 3.2: A tropical line.

Notice that different polytopes can play the role of Newton polytope for the same tropical hypersurface in $\mathbb{R}^n$. For example, given a finite collection A of integer points in $\mathbb{R}^n$, a function $\varphi : A \to \mathbb{R}$, and any integer point c in $\mathbb{R}^n$, consider the pair (A', φ'), where $A' = A + c$ and $\varphi : A' \to \mathbb{R}$ is the function such that $\varphi'(x) = \varphi(x - c)$ for any point $x \in A'$. Then, the tropical hypersurfaces $T(A, \varphi)$ and $T(A', \varphi')$ coincide. This example is a typical one: any two Newton polytopes of a given tropical hypersurface in $\mathbb{R}^n$ differ by a translation by an integer vector.

Consider a tropical hypersurface $T(A, \varphi) \subset \mathbb{R}^n$ associated with a pair (A, φ). The function φ gives rise to a subdivision of the convex hull $\Delta(A)$ of A. Namely, let $F : \Delta(A) \to \mathbb{R}$ be the convex piecewise-linear function whose graph is the lower part of the convex hull of the graph of φ. The linearity domains of F are n-dimensional polytopes with integer vertices. These polytopes produce a subdivision $S(A, \varphi)$ of $\Delta(A)$.

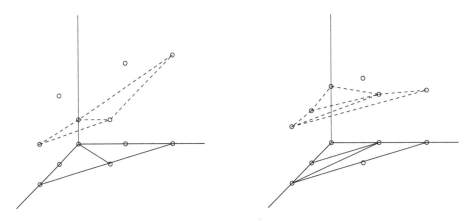

Figure 3.3: Examples of subdivisions of the triangle with vertices $(0,0)$, $(2,0)$, and $(0,2)$.

The subdivision $S(A, \varphi)$ is dual to the subdivision $\Theta(A, \varphi)$ in the following sense.

Theorem 3.3 (Duality theorem). *There exists a one-to-one correspondence $\mathcal{B}$ between the elements of $S(A, \varphi)$ on one side and the elements of $\Theta(A, \varphi)$ on the other side such that*

- *if e is an element of $S(A, \varphi)$ having dimension i, then the element $\mathcal{B}(e)$ of $\Theta(A, \varphi)$ has dimension $n - i$, and the linear spans of e and $\mathcal{B}(e)$ are orthogonal,*

- *the correspondence $\mathcal{B}$ reverses the incidence relation.*

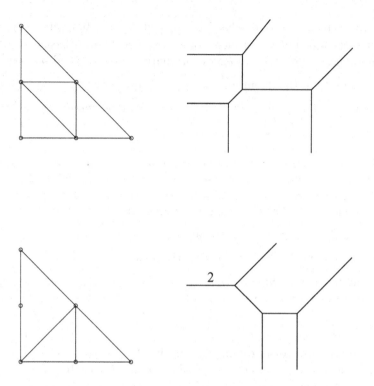

Figure 3.4: Examples of tropical conics in $\mathbb{R}^2$ and corresponding dual subdivisions (the weight of an edge is indicated only if this weight is different from 1).

3.3 Geometric description of plane tropical curves

We will now restrict ourselves to the study of tropical curves in $\mathbb{R}^2$. These curves can be described in the following geometric way.

Let V be a finite collection of distinct points in $\mathbb{R}^2$, E_b a collection of segments whose endpoints belong to V, and E_u a finite collection of half-infinite rays whose endpoints belong to V. Assume that the intersection of any two elements in $E_b \cup E_u$ is either a point in V or empty. Let $w : E_b \cup E_u \rightarrow \mathbb{N}$ be a function (for an element $e \in E_b \cup E_u$, the number $w(e)$ is called the *weight* of e). Such a quadruple (V, E_b, E_u, w) is called a *weighted rectilinear graph*. A weighted rectilinear graph (V, E_b, E_u, w) is *balanced* if

- each element in $E_b \cup E_u$ has a rational slope,

- no element in V is adjacent to exactly two elements in $E_b \cup E_u$,

- for any element v in V, one has $\sum_{e_k \in E(v)} w(e_k) \cdot \vec{e_k} = 0$, where $E(v) \subset E_b \cup E_u$ is the set formed by the elements of $E_b \cup E_u$ which are adjacent to v, and $\vec{e_i}$ is the smallest vector with integer coordinates based at v and pointed along e_i.

The last property in the definition above is called the *balancing condition*.

Theorem 3.4. *Any tropical curve in $\mathbb{R}^2$ represents a balanced weighted rectilinear graph. Conversely, any balanced weighted rectilinear graph represents a tropical curve.*

Proof. Let $T(A, \varphi)$ be a tropical curve associated with a pair (A, φ). Consider the weighted rectilinear graph Γ whose set V (respectively, E_b, E_u) is formed by the vertices (respectively, bounded edges, unbounded edges) of $T(A, \varphi)$, and whose weights coincide with the corresponding weights of $T(A, \varphi)$. Since

- any edge in $S(A, \varphi)$ has a rational slope,

- any polygon in $S(A, \varphi)$ has at least three sides,

- for any polygon in $S(A, \varphi)$ with vertices $p_1, p_2, \ldots, p_n$, one has $\overrightarrow{p_1 p_2} + \cdots + \overrightarrow{p_{n-1} p_n} + \overrightarrow{p_n p_1} = 0$,

the Duality Theorem 3.3 implies that the graph Γ verifies all three properties appearing in the definition of balanced graphs, and thus, is balanced.

To prove the converse statement, consider a balanced weighted graph Γ, and choose a region R of the complement of Γ in $\mathbb{R}^2$. Associate to R an arbitrary affine-linear function $\widehat{\varphi}_R : \mathbb{R}^2 \to \mathbb{R}$, $\widehat{\varphi}_R(x, y) = i_R x + j_R y - \varphi_R$. Let R' be a region neighboring to R, i.e., such that the intersection e of the closures of R and R' is an element of $E_b \cup E_u$. Associate to R' the affine-linear function $\widehat{\varphi}_{R'} : \mathbb{R}^2 \to \mathbb{R}$, $\widehat{\varphi}_{R'}(x, y) = i_{R'} x + j_{R'} y - \varphi_{R'}$ such that $((i_{R'} - i_R)/w(e), (j_{R'} - j_R)/w(e))$ is the smallest integer vector normal to e and pointed inside R', and the restrictions of $\widehat{\varphi}_R$ and $\widehat{\varphi}_{R'}$ on e coincide. Continuing in the same manner, we associate to any region P of the complement of Γ in $\mathbb{R}^2$ an affine-linear function $\widehat{\varphi}_P : \mathbb{R}^2 \to \mathbb{R}$, $\widehat{\varphi}(x, y) = i_P x + j_P y - \varphi_P$. The balancing condition insures that the function $\widehat{\varphi}_P$ does not depend on the sequence of regions used in the definition of $\widehat{\varphi}_P$. We obtain a finite collection A of integer points (i_P, j_P) and a function $\varphi : A \to \mathbb{R}$ defined by $(i_P, j_P) \mapsto \varphi_P$ such that Γ represents the tropical curve associated with (A, φ). $\square$

The *sum* $T(A_1, \varphi_1) + \cdots + T(A_n, \varphi_n)$ of plane tropical curves $T(A_1, \varphi_1), \ldots, T(A_n, \varphi_n)$ is the plane tropical curve defined by the tropical polynomial $\widehat{\varphi}_1 + \cdots + \widehat{\varphi}_n$. The underlying set of the tropical curve $T(A_1, \varphi_1) + \cdots + T(A_n, \varphi_n)$ is the union of underlying sets of $T(A_1, \varphi_1), \ldots, T(A_n, \varphi_n)$, and the weight of any edge of $T(A_1, \varphi_1) + \cdots + T(A_n, \varphi_n)$ is equal to the sum of the weights of the corresponding edges of summands. A tropical curve in $\mathbb{R}^2$ is *reducible* if it is the sum of two proper tropical subcurves. A non-reducible tropical curve in $\mathbb{R}^2$ is called *irreducible*.

Tropical curves have many properties in common with algebraic curves. For example, one can prove the following analog of the Bézout theorem (see, e.g., [64]).

Theorem 3.5 (Tropical Bézout theorem). *Let T_1 and T_2 be two tropical curves of degrees m_1 and m_2, respectively, such that T_1 and T_2 are in general position with respect to each other (the latter condition means that T_1 and T_2 intersect each other only in inner points of edges); then the number of intersection points (counted with certain multiplicities) of T_1 and T_2 is equal to $m_1 m_2$. The multiplicities of intersection points are defined as follows. Consider an intersection point of an edge e_1 of T_1 and an edge e_2 of T_2. Let (a_1, b_1) and (a_2, b_2) be smallest integer vectors along e_1 and e_2, respectively. Then, the multiplicity of the intersection point is equal to $w(e_1) w(e_2) |a_1 b_2 - a_2 b_1|$.*

Notice also that for any two points in general position in $\mathbb{R}^2$ there exists exactly one tropical line passing through these points. This observation has an important generalization which is the core of the remaining part of this chapter.

3.4 Count of complex nodal curves

In this section, we formulate certain enumerative problems concerning nodal curves in the complex projective plane $\mathbb{C}P^2$.

Fix a positive integer m, and choose $\frac{m(m+3)}{2}$ points in general position in $\mathbb{C}P^2$. There exists exactly one curve of degree m in $\mathbb{C}P^2$ which passes through the chosen points. Indeed, the space $\mathbb{C}C_m$ of all the curves of degree m in $\mathbb{C}P^2$ can be identified with a complex projective space $\mathbb{C}P^N$ of dimension $N = \frac{m \cdot (m+3)}{2}$: the coefficients of a polynomial defining a given curve can be taken for homogeneous coordinates of the corresponding point in $\mathbb{C}C_m$. The condition to pass through a given point in $\mathbb{C}P^2$ produces a linear equation in coefficients of a polynomial defining the curve, and thus determines a hyperplane in $\mathbb{C}C_m$. If the configuration of $\frac{m(m+3)}{2}$ chosen points is sufficiently generic, the corresponding $\frac{m(m+3)}{2}$ hyperplanes in $\mathbb{C}C_m$ have exactly one common point (and in addition this point corresponds to a nonsingular curve).

Choose now $\frac{m \cdot (m+3)}{2} - 1$ points in general position in $\mathbb{C}P^2$. How many curves of degree m with one non-degenerate double point each pass through the chosen points? Consider the hypersurface $D \subset \mathbb{C}C_m$ formed by the points corresponding to singular curves. The hypersurface D is called the *discriminant* of $\mathbb{C}C_m$. The smooth part of D is formed by the points corresponding to curves whose only singular point is non-degenerate double. If the configuration of $\frac{m(m+3)}{2} - 1$ chosen points is sufficiently generic, the intersection of the hyperplanes corresponding to these points is a line in $\mathbb{C}C_m$, and moreover, this line intersects the discriminant only in the smooth part and transversally. Thus, the number we are interested in is the degree of D.

Consider the following generalization of the above questions. Pick an integer δ verifying the inequalities $0 \leq \delta \leq \frac{(m-1)(m-2)}{2}$, and choose a collection U of $\frac{m(m+3)}{2} - \delta$ points in $\mathbb{C}P^2$. Consider the curves of degree m in $\mathbb{C}P^2$ which pass through all the points of U and have δ non-degenerate double points each. If U is sufficiently generic, then the number of these curves is finite and does not depend on U. Denote by $N_m(\delta)$ (respectively, $N_m^{\mathrm{irr}}(\delta)$) the number of curves (respectively, of irreducible curves) of degree m in $\mathbb{C}P^2$ which pass through the points of a generic configuration of $\frac{m(m+3)}{2} - \delta$ points in $\mathbb{C}P^2$ and have δ non-degenerate double points each. The expression "sufficiently generic" in the description of the numbers $N_m(\delta)$ can be made precise in the following way. Denote by $S_m(\delta)$ the subset of $\mathbb{C}C_m$ formed by the points corresponding to curves of degree m having δ non-degenerate double points each and no other singularities. The *Severy variety* $\overline{S}_m(\delta)$ is the closure of $S_m(\delta)$ in $\mathbb{C}C_m$. It is an algebraic variety of codimension δ in $\mathbb{C}C_m$. Its smooth part is $S_m(\delta)$. We say that a collection U of $\frac{m(m+3)}{2} - \delta$ points is *generic*, if the dimension of the projective subspace $\Pi(U) \subset \mathbb{C}C_m$ defined by the points of U is equal to δ, the intersection $\Pi(U) \cap \overline{S}_m(\delta)$ is contained in $S_m(\delta)$, and this intersection is transverse. Generic collections form an open dense subset in the space of all collections of $\frac{m(m+3)}{2} - \delta$ points in $\mathbb{C}P^2$. If U is generic, the number of curves of degree m in $\mathbb{C}P^2$ which pass through the points of U and have δ non-degenerate double points each is equal to the number of elements in the finite intersection $\Pi(U) \cap \overline{S}_m(\delta)$. Thus, the number $N_m(\delta)$ is the degree of the Severi variety $\overline{S}_m(\delta)$.

The numbers $N_m(\delta)$ can be calculated starting with the numbers $N_m^{\mathrm{irr}}(\delta)$ and *vice versa* (see, for example, [4]). The numbers $N_m^{\mathrm{irr}}(\delta)$ are *Gromov–Witten invariants* of $\mathbb{C}P^2$. The number $N_m^{\mathrm{irr}}(\delta)$, where $\delta = \frac{(m-1)(m-2)}{2}$, is the number of rational curves of degree m which pass through a generic collection of $\frac{m(m+3)}{2} - \frac{(m-1)(m-2)}{2} = 3m - 1$ points in $\mathbb{C}P^2$. A recursive formula for the numbers $N_m^{\mathrm{irr}}(\delta)$, with $\delta = \frac{(m-1)(m-2)}{2}$, was found by M. Kontsevich (see [33]). A recursive formula that allows one to calculate the numbers $N_m(\delta)$ with an arbitrary δ was obtained by L. Caporaso and J. Harris [4].

G. Mikhalkin proposed a new formula for the numbers $N_m(\delta)$ (see [40, 42]). This formula has an immediate generalization to the case of an arbitrary toric surface (see [40, 42]). Mikhalkin's approach is based on a reformulation of the enumerative problem presented above into an enumerative problem concerning tropical curves.

3.5 Correspondence theorem

To formulate Mikhalkin's correspondence theorem, introduce additional definitions.

Let m be a positive integer, $\Delta_m \subset \mathbb{R}^2$ the triangle having the vertices $(0,0)$, $(m,0)$, and $(0,m)$, and T a tropical curve of degree m. The curve T is called *simple* if the corresponding dual subdivision S_T of Δ_m satisfies the following properties:

- any polygon of S_T is either a triangle or a parallelogram,

- any integer point on the boundary of Δ_m is a vertex of S_T.

In this case, the subdivision S_T is also called simple. Notice that if T is simple, then it can be represented in a unique possible way as a sum of irreducible tropical curves.

Assume that T is simple. Then, the *rank* of T is the difference diminished by 1 between the number of vertices of S_T and the number of parallelograms in S_T. The *multiplicity* $\mu(S_T)$ of S_T (and the multiplicity $\mu(T)$ of T) is the product of areas of all the triangles in S_T (we normalize the area in such a way that the area of a triangle whose only integer points are its vertices is equal to 1).

Let r be a positive integer, and $\mathcal{U}$ a generic collection of r points in $\mathbb{R}^2$. (One can formalize the expression "generic" used here and introduce the notion of a *tropically generic* collection of points in $\mathbb{R}^2$; this can be done in a way similar to the one used in the complex situation.) Consider the collection $\mathcal{C}(\mathcal{U})$ of simple tropical curves of degree m and of rank r which pass through all the points of $\mathcal{U}$. Denote by $\mathcal{C}^{\mathrm{irr}}(\mathcal{U})$ the collection of irreducible curves belonging to $\mathcal{C}(\mathcal{U})$.

Theorem 3.6. (G. Mikhalkin, [42]). *Let $\mathcal{U}$ be a generic set of* $r = \frac{m(m+3)}{2} - \delta$ *points in $\mathbb{R}^2$, where an integer δ satisfies the inequalities $0 \leq \delta \leq \frac{(m-1)(m-2)}{2}$. Then,*

$$N_m(\delta) = \sum_{T \in \mathcal{C}(\mathcal{U})} \mu(T) \quad \text{and} \quad N_m^{\mathrm{irr}}(\delta) = \sum_{T \in \mathcal{C}^{\mathrm{irr}}(\mathcal{U})} \mu(T).$$

Theorem 3.6 is a particular case of Mikhalkin's theorem which is valid in the more general setting of projective toric surfaces. Mikhalkin's proof of Theorem 3.6 provides a bijection between the multi-set $\mathcal{C}(\mathcal{U})$ and the set of complex curves of degree m which pass through certain r generic points in $\mathbb{C}P^2$ and have δ non-degenerate double points each. A slightly different approach establishing such a bijection was proposed by E. Shustin [59].

In addition, Mikhalkin [40, 42] found a combinatorial algorithm which gives a possibility to calculate the number of tropical curves in question. We present this algorithm in the next section.

3.6 Mikhalkin's algorithm

Let again m be a positive integer, and Δ_m the triangle with vertices $(0,0)$, $(m,0)$, and $(0,m)$. Fix a linear function $\lambda : \mathbb{R}^2 \to \mathbb{R}$ which is injective on the integer points of Δ_m, and denote by p (respectively, q) the vertex of Δ_m where λ takes its minimum (respectively, maximum). The points p and q divide the boundary of Δ_m in two parts. Denote one of these parts by $\partial\Delta_+$, and the other part by $\partial\Delta_-$.

Let l be a natural number. A path $\gamma : [0, l] \to \Delta_m$ is called λ-*admissible* if

- $\gamma(0) = p$ and $\gamma(l) = q$,

- the composition $\lambda \circ \gamma$ is injective,

- for any integer $0 \leq i \leq l - 1$ the point $\gamma(i)$ is integer, and $\gamma([i, i + 1])$ is a segment.

The number l is called the *length* of γ, and the integer points of the form $\gamma(i)$, where i is an integer satisfying the inequalities $0 \leq i \leq l$, are called *vertices* of γ. A λ-admissible path γ divides Δ_m in two parts: the part $\Delta_+(\gamma)$ bounded by γ and $\partial\Delta_+$ and the part $\Delta_-(\gamma)$ bounded by γ and $\partial\Delta_-$. Define an operation of *compression* of $\Delta_+(\gamma)$ in the following way. Let j be the smallest positive integer $1 \leq j \leq l-1$ such that $\gamma(j)$ is the vertex of $\Delta_+(\gamma)$ with the angle less than π (a compression of $\Delta_+(\gamma)$ is defined only if such an integer j does exist). A compression of $\Delta_+(\gamma)$ is $\Delta_+(\gamma')$, where γ' is either the path defined by $\gamma'(i) = \gamma(i)$ for $i < j$ and $\gamma'(i) = \gamma(i+1)$ for $i \geq j$, or the path defined by $\gamma'(i) = \gamma(i)$ for $i \neq j$ and $\gamma'(j) = \gamma(j-1) + \gamma(j+1) - \gamma(j)$ (the latter path can be considered only if $\gamma(j-1) + \gamma(j-1) + \gamma(j) \in \Delta_m$). Note that γ' is also a λ-admissible path. A sequence of compressions started with $\Delta_+(\gamma)$ and ended with a path whose image coincides with $\partial\Delta_+$ defines a subdivision of $\Delta_+(\gamma)$ which is called *compressing*. A compression and a compressing subdivision of $\Delta_-(\gamma)$ is defined in a completely similar way. A pair $(S_+(\gamma), S_-(\gamma))$, where $S_\pm(\gamma)$ is a compressing subdivision of $\Delta_\pm(\gamma)$, produces a subdivision of Δ_m. The latter subdivision is called γ-*consistent*. Denote by $\mathcal{N}_\lambda(\gamma)$ the collection of simple γ-consistent subdivisions of Δ_m.

Theorem 3.7. (G. Mikhalkin, see [40, 42]). *Let* $0 \leq \delta \leq \frac{(m-1)(m-2)}{2}$ *be an integer. There exists a generic set* $\mathcal{U}$ *of* $r = \frac{m(m+3)}{2} - \delta$ *points in* $\mathbb{R}^2$ *such that the map associating to a simple tropical curve* T *of degree* m *the dual subdivision* S_T *of* Δ_m *establishes a one-to-one correspondence between the set* $\mathcal{C}(\mathcal{U})$ *and the disjoint union* $\amalg_\gamma \mathcal{N}_\lambda(\gamma)$, *where* γ *runs over all the* λ-*admissible paths in* Δ_m *of length* r. *In particular,* $N_m(\delta) = \sum_\gamma \sum_{S \in \mathcal{N}_\lambda(\gamma)} \mu(S)$, *where* $\mu(S)$ *is the multiplicity of* S.

Example 3.8. Figure 3.5 illustrates the algorithm in the case of rational cubics. The function λ is given by $\lambda(i, j) = i - \varepsilon j$, where ε is a positive sufficiently small number. In this case, the number of integer points of the Newton triangle Δ_3 is greater by 1 than the number $r+1 = 9$ of vertices of λ-admissible paths to consider. Therefore, each λ-admissible path γ of length $r = 8$ is uniquely determined by the integer point which is not a vertex of γ. It is easy to see that the integer points marked by small squares on Figure 3.5 are vertices of any path γ such that the set of simple γ-consistent subdivisions is not empty.

3.7 Welschinger invariants

Mikhalkin's correspondence theorem also gives a possibility to enumerate real curves passing through specific configurations of real points in $\mathbb{R}P^2$ (as well as on other projective toric surfaces). Of course, in the real case the result depends on the chosen point configuration in $\mathbb{R}P^2$. Fortunately, another important discovery was made recently by J.-Y. Welschinger [73, 74]. He found a way of attributing weights ± 1 to real rational curves which makes the number of curves counted with the weights to be independent of the configuration of points in $\mathbb{R}P^2$ and produces lower bounds for the number of real curves in question.

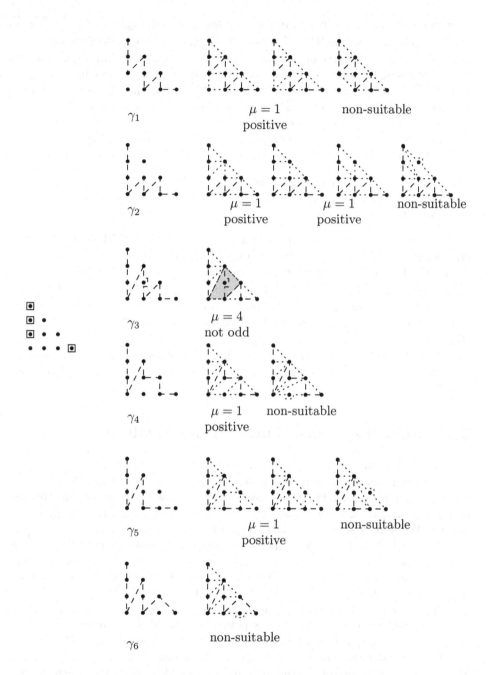

Figure 3.5: The algorithm for rational cubics.

For a given positive integer m and an integer δ satisfying $0 \leq \delta \leq \frac{(m-1)(m-2)}{2}$, choose a generic collection U of $\frac{m(m+3)}{2} - \delta$ points in $\mathbb{R}P^2$. Consider the set of real irreducible curves of degree m passing through all the points of U and having δ non-degenerate double points each. Denote by $R_m^{\mathrm{irr}}(\delta, U)$ the number of curves in the set considered, and by $R_m^{\mathrm{irr, \, even}}(\delta, U)$ (resp., $R_m^{\mathrm{irr, \, odd}}(\delta, U)$) the number of curves in this set which have even (resp., odd) number of solitary nodes (i.e., non-degenerate real double points locally given by the equation $x^2 + y^2 = 0$). The *Welschinger sign* of a nodal real curve is $(-1)^s$, where s is the number of solitary nodes of the curve. Define the *Welschinger number* as $W_m(\delta, U) = R_m^{\mathrm{irr, \, even}}(\delta, U) - R_m^{\mathrm{irr, \, odd}}(\delta, U)$.

Theorem 3.9. (J.-Y. Welschinger, see [73, 74]). *If $\delta = \frac{(m-1)(m-2)}{2}$ (i.e., if the considered curves are rational), then $W_m(\delta, U)$ does not depend on the choice of a (generic) configuration U.*

In fact, Theorem 3.9 is a particular case of Welschinger's theorem. The general statement in the case of real symplectic 4-manifolds and the proof can be found in [73, 74]. Higher dimensional generalizations are found in [75].

The number $W_m(\frac{(m-1)(m-2)}{2}, U)$ is called the *Welschinger invariant* and is denoted by W_m. Clearly, the Welschinger invariant W_m gives a lower bound for the number of real solutions to our interpolation problem: $R_m^{\mathrm{irr}}(\frac{(m-1)(m-2)}{2}, U) \geq |W_m|$.

Welschinger's theorem provides another type of applications of Mikhalkin's correspondence. The remaining part of this chapter is mostly devoted to these applications.

3.8 Welschinger invariants W_m for small m

Let us calculate the Welschinger invariants W_m for $m = 1$, 2, and 3.

If $m = 1$, then we should count straight lines passing through $3 \cdot 1 - 1 = 2$ points in general position in $\mathbb{R}P^2$. There is exactly one straight line passing through two points in general position in $\mathbb{R}P^2$. This is a nonsingular real rational curve, and its Welschinger sign is $+1$. Thus, $W_1 = 1$.

If $m = 2$, then we should count real conics passing through $3 \cdot 2 - 1 = 5$ points in general position in $\mathbb{R}P^2$. There is exactly one curve of degree 2 passing through five points in general position in $\mathbb{R}P^2$. Once again, this curve is real, rational and nonsingular, its Welschinger sign is $+1$. Thus, $W_2 = 1$.

The case $m = 3$ is more complicated. If $m = 3$, we should count real rational cubics passing through $3 \cdot 3 - 1 = 8$ points in general position in $\mathbb{R}P^2$. Let U be a generic configuration of 8 points in $\mathbb{R}P^2$. The configuration U defines a pencil $\mathcal{P}$ of real cubics passing through all the points of U. Any two cubics of $\mathcal{P}$ intersect in 8 points of U, and thus have one additional point of intersection in $\mathbb{C}P^2$. Denote this point by Q. Notice that Q is real, and all the cubics of $\mathcal{P}$ pass through Q. Let $\widetilde{\mathbb{R}P^2}$ be $\mathbb{R}P^2$ blown up at 8 points of U and at the point Q.

The Euler characteristic $\chi(\widetilde{\mathbb{R}P^2})$ of $\widetilde{\mathbb{R}P^2}$ is equal to $1 - 9 = -8$. On the other hand, the calculation of the Euler characteristic of $\widetilde{\mathbb{R}P^2}$ via the pencil $\mathcal{P}$ gives $\chi(\widetilde{\mathbb{R}P^2}) = R_3^{\text{irr, odd}}(1, U) - R_3^{\text{irr, even}}(1, U) = -W_3$. Thus, $W_3 = 8$.

The lower bound 8 for the number of real rational cubics passing through 8 points in general position in $\mathbb{R}P^2$ is sharp and was proved by V. Kharlamov before the discovery of the Welschinger invariants (see, for example, [7]). It is not known whether the lower bounds provided by the Welschinger invariants W_m, $m \geq 4$ are sharp.

3.9 Tropical calculation of Welschinger invariants

A simple tropical curve T of degree m and the corresponding subdivision S_T of Δ_m are called *odd*, if each triangle in S_T has an odd (normalized) area. Such a curve T and the dual subdivision S_T are called *positive* (respectively, *negative*) if the sum of the numbers of interior integer points over all the triangles of S_T is even (respectively, odd). Associate to any simple tropical curve T of degree m and to the corresponding subdivision S_T of Δ_m the Welschinger multiplicity $\mathcal{W}(T)$ in the following way. If T is not odd, then put $\mathcal{W}(T) = 0$. If T is odd and positive (respectively, negative), then put $\mathcal{W}(T) = 1$ (respectively, $\mathcal{W}(T) = -1$).

Theorem 3.10. (cf. [40, 42], and [59]). *Let $\mathcal{U}$ be a generic collection of $r = 3m - 1$ points in $\mathbb{R}^2$. Then,*

$$W_m = \sum_{T \in \mathcal{C}^{\text{irr}}(\mathcal{U})} \mathcal{W}(T).$$

Take now a set $\mathcal{U}$ with the properties described in Theorem 3.7, and denote by $n_\lambda^+(\gamma)$ (respectively, $n_\lambda^-(\gamma)$) the number of odd positive (respectively, negative) subdivisions in $\mathcal{N}_\lambda(\gamma)$ which are dual to irreducible tropical curves. The following statement is an immediate corollary of Theorems 3.10 and 3.7.

Theorem 3.11. (see [40, 42]). *The Welschinger invariant W_m is equal to $\sum_\gamma (n_\lambda^+(\gamma) - n_\lambda^-(\gamma))$, where γ runs over all the λ-admissible paths in Δ_m of length $r = 3m - 1$.*

Figure 3.5 illustrates the tropical calculation of W_3. The subdivision with two grey triangles has multiplicity 4. This subdivision is not odd (the grey triangles are of area 2), and thus, it does not contribute to W_3.

Remark 3.12. As it was noticed by G. Mikhalkin, one can easily prove the following result comparing Theorems 3.7 and 3.11: *for any positive integer m, the Welschinger invariant W_m and the corresponding Gromov–Witten invariant $N_m = N_m^{\text{irr}}(\frac{(m-1)(m-2)}{2})$ are congruent modulo 4.*

Theorem 3.11 gives a possibility to calculate or to estimate the Welschinger invariants W_m. The following section is devoted to applications of Theorem 3.11.

3.10 Asymptotic enumeration of real rational curves

Consider the following question: *fix a positive integer m; whether for any generic collection of $3m - 1$ points in the real projective plane there always exists a real rational curve of degree m which passes through the points of the collection ?* (The number $N_m = N_m^{\mathrm{irr}}(\frac{(m-1)(m-2)}{2})$ of complex rational curves (see [33]) is even for every $m \geq 3$, so the existence of a required real curve does not immediately follow from the computation in the complex case.)

The following statement is a corollary of Theorem 3.11.

Theorem 3.13. (I. Itenberg, V. Kharlamov, E. Shustin; see [27, 28]))

- *For any positive integer m, the Welschinger invariant W_m is positive.*

- *The sequences $\log W_m$ and $\log N_m$, $m \in \mathbb{N}$, are asymptotically equivalent, More precisely,*

$$\log W_m = \log N_m + O(m) \quad and$$
$$\log N_m = 3m \log m + O(m).$$

As a corollary, the aforementioned question is answered in the affirmative. Moreover, Theorems 3.9 and 3.13 imply that asymptotically in the logarithmic scale all the complex solutions of our interpolation problem are real.

Let $\lambda^0 : \mathbb{R}^2 \to \mathbb{R}$ be a linear function defined by $\lambda^0(i, j) = i - \varepsilon j$, where ε is a sufficiently small positive number (so that λ^0 defines a kind of a lexicographical order on the integer points of the triangle Δ_m).

The following statement is a key point in the proof of Theorem 3.13.

Lemma 3.14. *For any λ^0-admissible path γ in Δ_m, the number $n_{\lambda^0}^-(\gamma)$ is equal to 0.*

Proof. Let γ be a λ^0-admissible path in Δ_m, and S a subdivision in the collection $\mathcal{N}_{\lambda^0}(\gamma)$. The subdivision S does not have an edge with the endpoints (i_1, j_1) and (i_2, j_2) such that $|i_1 - i_2| > 1$; otherwise, at least one integer point on the boundary of Δ_m would not be a vertex of the corresponding compressing subdivision. This implies that no triangle in S has interior integer points. □

Lemma 3.14 implies that, for any λ_0-admissible path γ, the contribution of any γ-consistent subdivision of Δ_m to the Welschinger invariant W_m is non-negative. Thus, to prove Theorem 3.13, it is sufficient to present a λ_0-admissible path γ such that the contribution of certain γ-consistent subdivisions of Δ_m to the Welschinger invariant W_m is big enough.

Sketch of the proof of Theorem 3.13. Inscribe in Δ_m a sequence of maximal size squares as shown on Figure 3.6(a). Their right upper vertices have the coordinates

$$(x_i, y_i), \ i \geq 1, \ x_1 = y_1 = \left[\frac{m}{2}\right], \ y_{i+1} = \left[\frac{m - x_i}{2}\right], \ x_{i+1} = x_i + y_{i+1}.$$

Put $(x_0, y_0) = (0, m)$. Then pick a λ^0-admissible path γ consisting of segments of integer length 1 as shown on Figure 3.6(b). This path consists of sequences of vertical segments joining (x_i, y_i) with $(x_i, y_{i+1} - 1)$, zig-zag sequences joining $(x_i, y_{i+1} - 1)$ with (x_{i+1}, y_{i+1}) (in such a zig-zag sequence the segments of slope 1 alternate with vertical segments; it always starts and ends with segments of slope 1), and the segments $[(m - 1, 1), (m - 1, 0)]$ and $[(m - 1, 0), (m, 0)]$. The length of this path is $3m - 1$.

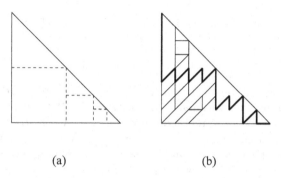

(a)   (b)

Figure 3.6: Path γ and γ-consistent subdivisions of Δ_m.

Now, we select some γ-consistent subdivisions of Δ_m. Subdivide the upper part $\Delta_+(\gamma)$ in vertical strips of integer width 1. Note that the rightmost strip consists of one primitive triangle (a triangle with integer vertices is called *primitive* if it is of (normalized) area 1). Pack into each strip but the rightmost one the maximal possible number of primitive parallelograms (a parallelogram with integer vertices is called primitive if it is of (normalized) area 2) and place in the remaining part of the strip two primitive triangles (see Figure 3.6(b)). Then subdivide $\Delta_-(\gamma)$ in slanted strips of slope 1 and horizontal width 1. Pack into each strip the maximal possible number of primitive parallelograms. This gives a subdivision of any slanted strip situated above the line $y = x$. For any strip situated below the line $y = x$ place in the remaining part of the strip one primitive triangle (see Figure 3.6(b)). The total number of such γ-consistent subdivisions is

$$M_m \geq \prod_i \frac{y_i!(y_i + 1)!}{2^{y_i}} \cdot \prod_i y_i! \,, \qquad (3.1)$$

where the first product corresponds to subdivisions of $\Delta_+(\gamma)$ and the second one to those of $\Delta_-(\gamma)$.

All the constructed subdivisions of Δ_m are simple and odd, each of them is dual to an irreducible tropical curve and contributes 1 to the Welschinger invariant. The irreducibility of the dual tropical curve can easily proved by the following induction. Let us scan the subdivision by vertical lines from right to left. The rightmost fragment of the tropical curve is dual to the primitive triangle $\Delta_m \cap \{x \geq m-1\}$, so it is irreducible. At the i-th step, $i > 0$, we look at the irreducible components of the curve dual to the union of those elements of our subdivision which intersect the strip $m-i-1 < x < m-i$. Each of these irreducible components either connects the lines $x = m - i - 1$ to $x = m - i$, or contains a pattern dual to a triangle with an edge on $x = m - i$, or contains a pattern dual to a slanted parallelogram. Therefore, each component joins the curve dual to the subdivision of $\Delta_m \cap \{x \geq m - i\}$.

We have $W_m \geq M_m$. This gives the first statement of Theorem 3.13, and since $\log N_m = 3m \log m + O(m)$, it remains to check that

$$\log M_m \geq 3m \log m + O(m). \qquad \square$$

Remark 3.15. The statements similar to Theorem 3.13 are proved for all unnodal (i.e., not containing any rational $(-n)$-curve, $n \geq 2$) toric Del Pezzo surfaces equipped with their standard real structure, see [28]. (A *real structure* on a complex variety X is an anti-holomorphic involution $conj : X \to X$. A subvariety $C \subset X$ is *real* with respect to $conj$ if $conj(C) = C$. The standard real structure on a toric variety is the one which is naturally compatible with the toric structure.) Recall that there are five unnodal toric Del Pezzo surfaces: $\mathbb{C}P^2$, $\mathbb{C}P^1 \times \mathbb{C}P^1$, and P_k, $k = 1, 2, 3$, where P_k is the projective plane $\mathbb{C}P^2$ blown up at k points in general position. The same asymptotic statements are also proved for all unnodal toric Del Pezzo surfaces equipped with any non-standard real structure except the standard real $\mathbb{C}P^1 \times \mathbb{C}P^1$ blown up at two imaginary conjugated points; see [61, 30]. Recently, E. Brugallé and G. Mikhalkin [3] proved the statements similar to Theorem 3.13 and Remark 3.12 for Welschinger invariants of $\mathbb{C}P^3$.

3.11　Recurrence formula for Welschinger invariants

As it is shown in [29], the Welschinger invariants W_m can also be calculated using a recurrence formula. This formula can be seen as a real analog of the Caporaso–Harris formula [4] for relative Gromov–Witten invariants of $\mathbb{C}P^2$.

Denote by $\mathcal{G}$ the semigroup of sequences $\alpha = (\alpha_1, \alpha_2, \ldots) \in \mathbb{Z}^\infty$ with non-negative terms and finite norm $\|\alpha\| = \sum_i \alpha_i$. Each element of $\mathcal{G}$ contains only finitely many non-zero terms, so in the description of concrete sequences we omit zero terms after the last non-zero one. The only exception concerns the zero element of $\mathcal{G}$ (the sequence with all the terms equal to zero). This element is denoted by (0). For an element α in $\mathcal{G}$, put $J\alpha = \sum_{i=1}^{\infty} (2i-1)\alpha_i$. Define in $\mathcal{G}$ the following natural partial order: if each term of a sequence α is greater than or equal to the corresponding term of a sequence β, then we say that α is greater than or equal to β and write $\alpha \geq \beta$. For two elements $\alpha = (\alpha_1, \alpha_2, \ldots)$ and $\beta = (\beta_1, \beta_2, \ldots)$ of $\mathcal{G}$ such that $\alpha \geq \beta$, the sequence $\alpha - \beta$, whose i-th term is equal to $\alpha_i - \beta_i$, is an element of $\mathcal{G}$. Denote by θ_k the element in $\mathcal{G}$ whose k-th term is equal to 1 and all the other terms are equal to 0.

For $\alpha, \alpha^{(1)}, \ldots, \alpha^{(s)} \in \mathcal{G}$ such that $\alpha \geq \alpha^{(1)} + \cdots + \alpha^{(s)}$ put

$$
\begin{pmatrix} \alpha \\ \alpha^{(1)}, \ldots, \alpha^{(s)} \end{pmatrix} = \prod_{i=1}^{\infty} \frac{\alpha_i!}{\alpha_i^{(1)}! \cdots \alpha_i^{(s)}! (\alpha_i - \sum_{k=1}^{s} \alpha_i^{(k)})!} .
$$

Theorem 3.16. (I. Itenberg, V. Kharlamov, E. Shustin; cf. [29])) *Consider the family of numbers $W_m(\alpha, \beta)$ (indexed by a positive integer m and sequences α, β in $\mathcal{G}$ such that $J\alpha + J\beta = m$) defined by the initial conditions $W_1((1), (0)) = W_1((0), (1)) = 1$ and the recurrence relation (valid for any $m \geq 2$)*

$$
W_m(\alpha, \beta) = \sum_{\substack{k \geq 1 \\ \beta_k > 0}} W_m(\alpha + \theta_k, \beta - \theta_k)
$$

$$
+ \sum \begin{pmatrix} \alpha \\ \alpha^{(1)}, \ldots, \alpha^{(s)} \end{pmatrix} \frac{n!}{n_1! \ldots n_s!} \prod_{i=1}^{s} \left(\begin{pmatrix} \beta^{(i)} \\ \widetilde{\beta}^{(i)} \end{pmatrix} W_{m^{(i)}}(\alpha^{(i)}, \beta^{(i)}) \right) , \qquad (3.2)
$$

where

$$
n = 2m + \|\beta\| - 2, \quad n_i = 2m^{(i)} + \|\beta^{(i)}\| - 1, \ i = 1, \ldots, s,
$$

and the latter sum in formula (3.2) is taken over all collections $(m^{(1)}, \ldots, m^{(s)})$, $(\alpha^{(1)}, \ldots, \alpha^{(s)})$, $(\beta^{(1)}, \ldots, \beta^{(s)})$, and $(\widetilde{\beta}^{(1)}, \ldots, \widetilde{\beta}^{(s)})$ considered up to simultaneous permutations and satisfying the relations

$$
m^{(i)} \in \mathbb{Z}, \quad m_i \geq 0, \quad \alpha^{(i)}, \beta^{(i)}, \widetilde{\beta}^{(i)} \in \mathcal{G}, \quad J\alpha^{(i)} + J\beta^{(i)} = m^{(i)}, \ i = 1, \ldots, s,
$$

$$
\sum_{i=1}^{s} m^{(i)} = m - 1, \quad \sum_{i=1}^{s} \alpha^{(i)} \leq \alpha, \quad \sum_{i=1}^{s} \beta^{(i)} = \beta + \sum_{i=1}^{s} \widetilde{\beta}^{(i)},
$$

$$
s = \|\sum_{i=1}^{s} \beta^{(i)} - \beta\|, \quad \|\widetilde{\beta}^{(i)}\| = 1, \quad \beta^{(i)} \geq \widetilde{\beta}^{(i)}, \ i = 1, \ldots, s .
$$

Then, for any positive integer m, we have $W_m((0), (m)) = W_m$.

The numbers $W_m(\alpha, \beta)$ are *tropical relative Welschinger invariants*. They can be interpreted as numbers of tropical curves subject to certain constraints and counted with appropriate multiplicities (see [29]). Theorem 3.16 is a particular case of Theorem 4 in [29]. The latter theorem deals with tropical analogs of curves of arbitrary genus on any unnodal toric Del Pezzo surface. The proof follows ideas of A. Gathmann and H. Markwig [13, 14] who suggested a tropical version of the Caporaso–Harris formula.

3.12 Welschinger invariants $W_{m,i}$

We end these lectures with a definition and some properties of the Welschinger invariants $W_{m,i}$.

Let m be a positive integer. Consider a configuration U of $3m - 1$ points in general position in $\mathbb{C}P^2$ such that U is *real*, that is invariant under the involution of complex conjugation c acting in $\mathbb{C}P^2$. If the configuration U contains a non-real point z, then U contains also the point $c(z)$. Denote by i the number of pairs of conjugated non-real points in U.

As in Section 3.7, consider the set of real rational curves of degree m passing through all the points of U. Denote by $R_m(U)$ the number of curves in the set considered, and by $R_m^{\mathrm{even}}(U)$ (resp., $R_m^{\mathrm{odd}}(U)$) the number of curves in this set which have an even (resp., odd) number of solitary nodes. Define the Welschinger number $W_{m,i}(U)$ as $R_m^{\mathrm{even}}(U) - R_m^{\mathrm{odd}}(U)$.

Theorem 3.17. (J.-Y. Welschinger [73, 74]). *The number $W_{m,i}(U)$ does not depend on the choice of a (generic) real configuration U provided that the number of pairs of conjugated non-real points in U is equal to i.*

The number $W_{m,i}(U)$ is also called Welschinger invariant and is denoted by $W_{m,i}$. Of course, $W_{m,0} = W_m$.

A calculation similar to that made in Section 3.8 shows that $W_{3,i} = 8 - 2i$ for any integer $0 \leq i \leq 4$. Notice that in this case the values $W_{3,i}$ are interpolated by a linear function. Whatever is the integer $0 \leq i \leq 4$, the number $R_{3,i}(U)$ of real rational cubics attains the value $W_{3,i}$ for a suitable generic configuration U.

To calculate the Welschinger invariants $W_{m,i}$ for quartics and quintics, one can use birational transformations and Welschinger's wall-crossing formula (see [74], Theorem 2.2) which expresses the first finite difference of the function $i \mapsto W_{m,i}$ as twice the Welschinger invariant of $\mathbb{C}P^2$ blown up at one real point. For quartics the answer is as follows (see [28] for details):

i	0	1	2	3	4	5
W	240	144	80	40	16	0

These values $W_{4,i}$ are interpolated by a polynomial of degree 3,

$$W_{4,i} = -\frac{4}{3}i(i-1)(i-2) + 16i(i-1) - 96i + 240.$$

For quintics the Welschinger invariants take the values

i	0	1	2	3	4	5	6	7
W	18264	9096	4272	1872	744	248	64	64

which are interpolated by a polynomial of degree 6,

$$\begin{aligned} W_{5,i} = {} & \tfrac{4}{45}i(i-1)(i-2)(i-3)(i-4)(i-5) \\ & - \tfrac{32}{15}i(i-1)(i-2)(i-3)(i-4) + 32i(i-1)(i-2)(i-3) \\ & - 320i(i-1)(i-2) + 2172i(i-1) - 9168i + 18264. \end{aligned}$$

In the cases $m = 3$, 4, and 5, the degree of the interpolating polynomials happens to be smaller than for a generic interpolation data, that is, smaller than $[\frac{3d-1}{2}]$. It is no more the case for any $m \geq 6$.

One of the facts known about the Welschinger invariants $W_{m,i}$ is the following theorem.

Theorem 3.18. (I. Itenberg, V. Kharlamov, E. Shustin; see [28])) *Let $m \geq 3$ be an integer, and i a non-negative integer such that $i \leq 3$. Then, the Welschinger invariant $W_{m,i}$ is positive. Moreover,*

$$W_{m,0} > W_{m,1} > W_{m,2}.$$

Furthermore, for a family of Welschinger invariants $(W_{m,i})_{m \in \mathbb{N}}$, $m \geq 3$, with a given $i \leq 3$, one has

$$\log W_{m,i} = \log N_m + O(m).$$

The proof of Theorem 3.18 is based on the tropical formulas obtained by E. Shustin [60].

Remark 3.19. Statements of the same nature as Theorem 3.18 are proved for all unnodal toric Del Pezzo surfaces equipped with their standard real structure; see [28]. In the case of $\mathbb{C}P^1 \times \mathbb{C}P^1$, the Welschinger invariants depend on three integers: the bi-degree (m_1, m_2) of the real rational curves under consideration and the number i of conjugated non-real points in a given configuration of points. In this case, one can improve the result of Theorem 3.18 and show that all the Welschinger invariants $W_{(m_1,m_2),i}$ (these invariants are defined if m_1 and m_2 are positive integers, and i is a non-negative integer such that $i < m_1 + m_2$) of $\mathbb{C}P^1 \times \mathbb{C}P^1$ equipped with the standard real structure $(z_1, z_2) \mapsto (\overline{z_1}, \overline{z_2})$ are positive; see [28].

3.13 Exercises

Exercise 3.1. Find a convex polygon $\Delta \subset \mathbb{R}^2$ with integer vertices and functions $\varphi_1, \varphi_2 : A \to \mathbb{R}$, where $A = \Delta \cap \mathbb{Z}^2$, such that the underlying sets of the tropical curves $T(A, \varphi_1)$ and $T(A, \varphi_2)$ coincide, but the subdivisions of Δ defined by φ_1 and φ_2 do not.

Exercise 3.2. Let $\Delta \subset \mathbb{R}^2$ be a convex polygon with integer vertices, and $\nu : \Delta \to \mathbb{R}$ a convex piecewise-linear function defining a primitive triangulation of Δ (i.e., a triangulation whose vertices are integer and whose triangles are primitive). Show that the tropical curve $T(A, \varphi)$, where $A = \Delta \cap \mathbb{Z}^2$ and $\varphi = \nu|_A$, is homotopy equivalent to a bouquet of n circles, n being the number of interior integer points of Δ.

Exercise 3.3. Let $A \subset \mathbb{Z}^2$ be a finite non-empty set, and $\varphi : A \to \mathbb{R}$ a function. For any $c \in \mathbb{Z}^2$, put $A' = A + c$ and consider the function $\varphi' : A' \to \mathbb{R}$ defined by $f'(x) = f(x - c)$. Prove that the tropical curves $T(A', f')$ and $T(A, f)$ coincide.

Exercise 3.4. Let $A = \{(0,0), (0,1), (1,0), (1,1)\} \subset \mathbb{Z}^2$. The tropical curves associated with the pairs of the form (A, φ), where $\varphi : A \to \mathbb{R}$ is a function, are called *tropical curves of bi-degree* $(1,1)$. Show that, for any three points in general position in $\mathbb{R}^2$, there exists precisely one tropical curve of bi-degree $(1,1)$ passing through these points.

Exercise 3.5. Let $A \subset \mathbb{Z}^2$ be a set of $n \geq 2$ points. Denote by $\mathcal{T}(A)$ the set of tropical curves associated with the pairs of the form (A, φ), where $\varphi : A \to \mathbb{R}$ is a function. Prove that, for any $n - 1$ points in general position in $\mathbb{R}^2$, there exists precisely one tropical curve $T \in \mathcal{T}(A)$ passing through these points.

Exercise 3.6. Show that any tropical hypersurface T in $\mathbb{R}^n$ is *balanced*, that is, for any $(n - 2)$-dimensional face σ of T,

$$\sum_{\delta \supset \sigma} w(\delta) \cdot e(\delta, \sigma) = 0 \ ,$$

where δ runs over all $(n - 1)$-dimensional faces containing σ, $w(\delta)$ is the weight of δ, and $e(\delta, \sigma)$ is the smallest integer inner normal vector of $\sigma \subset \delta$. Formulate the converse statement and prove it.

Exercise 3.7. Let A_1 and A_2 be two finite nonempty sets of integer points in $\mathbb{R}^2$. Denote by Δ_i the convex hull of A_i, $i = 1, 2$. Consider functions $\varphi_1 : A_1 \to \mathbb{R}$ and $\varphi_2 : A_2 \to \mathbb{R}$ such that the tropical curves $T(A_1, \varphi_1)$ and $T(A_2, \varphi_2)$ intersect each other only at interior points of their edges. Prove the *tropical Bernstein theorem*: the number of intersection points of $T(A_1, \varphi_1)$ and $T(A_2, \varphi_2)$, counted with the same multiplicities as those defined in the tropical Bézout theorem, is equal to the mixed area of Δ_1 and Δ_2, that is, to the Euclidean area of the Minkowski sum $\Delta_1 + \Delta_2$ diminished by the Euclidean areas of Δ_1 and Δ_2.

Exercise 3.8. Compute the Welschinger invariants

- for rational curves of bi-degree $(2,2)$ on $\mathbb{C}P^1 \times \mathbb{C}P^1$ equipped with the real structure $(z_1, z_2) \mapsto (\overline{z_1}, \overline{z_2})$,

- for rational curves of bi-degree $(2,2)$ on $\mathbb{C}P^1 \times \mathbb{C}P^1$ equipped with the real structure $(z_1, z_2) \mapsto (\overline{z_2}, \overline{z_1})$.

Exercise 3.9. Using Mikhalkin's algorithm, compute the number of uninodal curves (a curve is *uninodal* if its only singular point is non-degenerate double) of degree $m \geq 3$ which pass through given $(m^2 + 3m - 2)/2$ points in general position in $\mathbb{C}P^2$.

List of Figures

Bibliography

[1] G. M. Bergman, *The logarithmic-limit set of an algebraic variety.* Trans. AMS **157** (1971), 459–469.

[2] E. Brugallé, *Géométries énumératives complexe, réelle et tropicale.* In: Géométrie tropicale X-UPS 2008 (ed: P. Harnick, A. Plagne, and C. Sabbah), Ecole Polytechnique, 2008.

[3] E. Brugallé and G. Mikhalkin, *Enumeration of curves via floor diagrams*, C. R. Acad. Sci. Paris, Sér. I **345** (2007), 329–334.

[4] L. Caporaso and J. Harris, *Counting plane curves of any genus*, Invent. Math. **131** (1998), no. 2, 345–392.

[5] M. Develin, F. Santos, and B. Sturmfels, *On the rank of a tropical matrix.* Combinatorial and computational geometry, Math. Sci. Res. Inst. Publ., 52, Cambridge Univ. Press, Cambridge, 2005, pp. 213–242.

[6] R. Connelly and D. W. Henderson, *A convex 3-complex not simplicially isomorphic to a strictly convex complex.* Proc. Camb. Phil. Soc. **88** (1980), no. 2, 299–306.

[7] A. Degtyarev and V. Kharlamov, *Topological properties of real algebraic varieties: de côté de chez Rokhlin.* Russ. Math. Surveys **55** (2000), no. 4, 735–814.

[8] M. Einsiedler, M. Kapranov, and D. Lind, *Non-Archimedean amoebas and tropical varieties*, J. Reine Angew. Math. **601** (2006), 139–157.

[9] N. Eriksson, K. Ranestad, B. Sturmfels, and S. Sullivant, *Phylogenetic algebraic geometry.* Projective varieties with unexpected properties, Walther de Gruyter, Berlin, 2005, pp. 237–255.

[10] S. Fiedler-Le Touzé and S. Yu. Orevkov, *A flexible affine M-sextic non-realizable algebraically.* J. Algebraic Geom. **11** (2002), no. 2, 293–310.

[11] M. Forsberg, M. Passare, and A. Tsikh, *Laurent determinants and arrangements of hyperplane amoebas.* Advances in Math. **151** (2000), 45–70.

[12] W. Fulton, *Introduction to toric varieties.* Ann. Math. Studies 131, Princeton Univ. Press, Princeton N.J., 1993.

[13] A. Gathmann and H. Markwig, *The numbers of tropical plane curves through points in general position*, J. Reine Angew. Math. **602** (2007), 155–177.

[14] A. Gathmann and H. Markwig, *The Caporaso-Harris formula and plane relative Gromov-Witten invariants in tropical geometry*, Mathematische Annalen **338** (2007), 845–868.

[15] S. Gaubert, and Max Plus, *Methods and Applications of (max,+) linear Algebra*. In: STACS 97 (Lübeck). Lecture Notes in Comput. Sci., vol. 1200, pp. 261–282. Springer, Berlin, 1997.

[16] I. M. Gelfand, M. M. Kapranov, and A. V. Zelevinski, *Discriminants, resultants and multidimensional determinants*. Birkhäuser, Boston, 1994.

[17] M. Gromov, *Pseudo-holomorphic curves in symplectic manifolds*. Invent. Math. **82** (1985), 307–347.

[18] M. Gross and P. M. H. Wilson, *Large complex structure limits of K3 surfaces*. J. Differential Geom. **55** (2000), no. 3, 475–646.

[19] I. Itenberg, *Contre-exemples á la conjecture de Ragsdale*. C. R. Acad. Sci. Paris, Ser. 1. **317** (1993), 277–282.

[20] I. Itenberg, *Viro's method and T-curves*. Algorithms in algebraic geometry and applications (Santander, 1994), Progr. Math. 143, Birkhäuser, Basel, 1996, pp. 177–192.

[21] I. Itenberg, *Topology of real algebraic T-surfaces*. Revista Mat. Univ. Complutense Madrid **10** (1997), 131–152.

[22] I. Itenberg, *Amibes des variétés algébriques et denombrement de courbes [d'après G. Mikhalkin]*. Séminaire N. Bourbaki, 2002 - 2003, exp. n° 921, Astérisque **294** (2004), 335–361.

[23] I. Itenberg, *Introduction à la géométrie tropicale*. In: Géométrie tropicale X-UPS 2008 (ed: P. Harnick, A. Plagne, and C. Sabbah), Ecole Polytechnique, 2008.

[24] I. Itenberg and E. Shustin, *Singular points and limit cycles of planar polynomial vector fields*. Duke Math. J. **102** (2000), no. 1, 1–37.

[25] I. Itenberg and E. Shustin, *Viro theorem and topology of real and complex combinatorial hypersurfaces*. Israel J. Math. **133** (2003), 189–238.

[26] I. Itenberg and E. Shustin, *Combinatorial patchworking of real pseudo-holomorphic curves*. Turkish J. Math. **26** (2002), no. 1, 27–51.

[27] I. Itenberg, V. Kharlamov, and E. Shustin, *Welschinger invariants and enumeration of real rational curves*. Intern. Math. Res. Notices **49** (2003), 2639–2653.

[28] I. Itenberg, V. Kharlamov, and E. Shustin, *Logarithmic equivalence of Welschinger and Gromov-Witten invariants.* Uspekhi Mat. Nauk 59 (2004), no. 6, 85–110 (Russian; English translation: Russian Math. Surveys 59 (2004), no. 6, 1093–1116).

[29] I. Itenberg, V. Kharlamov, and E. Shustin, *A Caporaso-Harris type formula for Welschinger invariants of real toric Del Pezzo surfaces.* Preprint arXiv: math.AG/0608549. To appear in Commentarii Math. Helvetici.

[30] I. Itenberg, V. Kharlamov, and E. Shustin, *New cases of logarithmic equivalence of Welschinger and Gromov-Witten invariants,* Proc. Steklov Institute Math. **258** (2007), 65–73.

[31] M. Joswig, *Tropical Halfspaces.* Combinatorial and computational geometry, Math. Sci. Res. Inst. Publ., 52, Cambridge Univ. Press, Cambridge, 2005, pp. 409–431.

[32] V. N. Kolokoltsov and V. P. Maslov, *Idempotent Analysis and Applications,* Kluwer Acad. Publ., Dordrecht, The Netherlands, 1997.

[33] M. Kontsevich and Yu. Manin, *Gromov-Witten classes, quantum cohomology and enumerative geometry,* Commun. Math. Phys. **164** (1994), 525-562.

[34] M. Kontsevich and Y. Soibelman, *Homological mirror symmetry and torus fibrations.* Symplectic geometry and mirror symmetry (Seoul, 2000), World Sci. Publishing, River Edge, NJ, 2001, pp. 203–263.

[35] M. Kontsevich and Yu. Tschinkel, *Nonarchimedean Kähler geometry.* Preprint, 2002.

[36] G. L. Litvinov and V. P. Maslov, *The correspondence principle for idempotent calculus and some computer applications.* Idempotency (Bristol, 1994), Publ. Newton Inst., 11, Cambridge Univ. Press, Cambridge, 1998, pp. 420–443.

[37] V. P. Maslov, *New superposition principle for optimization problems.* Séminaire sur les Equations avec Dérivées Partielles 1985/6, Centre Mathématiques de l'Ecole Polythechnique, Palaiseau, 1986, exposé 24.

[38] G. Mikhalkin, *Real algebraic curves, the moment map and amoebas.* Ann. of Math. (2) **151** (2000), no. 1, 309–326.

[39] G. Mikhalkin, *Amoebas of algebraic varieties.* Preprint arXiv: math.AG/0108225.

[40] G. Mikhalkin, *Counting curves via the lattice paths in polygons.* C. R. Acad. Sci. Paris, Ser. 1. **336** (2003), no. 8, 629–634.

[41] G. Mikhalkin, *Amoebas of algebraic varieties and tropical geometry.* Different faces of geometry/Donaldson, S. (ed.) et al. Kluwer, NY, 2004, pp. 257–300.

[42] G. Mikhalkin, *Enumerative tropical algebraic geometry in* $\mathbb{R}^2$. J. Amer. Math. Soc. **18** (2005), 313–377.

[43] G. Mikhalkin, *Introduction to Tropical Geometry.* Notes from the IMPA lectures in Summer 2007, Preprint arXiv:0709.1049.

[44] G. Mikhalkin and H. Rullgård, *Amoebas of maximal area*, Intern. Math. Res. Notices **9** (2001), 441–451.

[45] G. Mikhalkin. *Tropical Geometry.* Preprint, available at http://www.math.toronto.edu/ mikha/book.ps.

[46] G. Mikhalkin. *Tropical Geometry and its applications.* Sanz-Solé, Marta (ed.) et al., Proceedings of the ICM, Madrid, Spain, August 22–30, 2006. Volume II: Invited lectures. Zürich, European Math. Soc., 2006, pp. 827–852.

[47] G. Mikhalkin and I. Zharkov, *Tropical curves, their Jacobians, and Θ-functions.* Preprint arXiv: math.AG/0612267.

[48] T. Nishinou and B. Siebert, *Toric degenerations of toric varieties and tropical curves*, Duke. Math. J. **135** (2006), no. 1, 1–51.

[49] S. Yu. Orevkov, *Link theory and oval arrangements of real algebraic curves.* Topology **38** (1999), no. 4, 779–810.

[50] L. Pachter and B. Sturmfels, *Tropical geometry of statistical models.* Proc. Natl. Acad. Sci. USA **101** (2004), no. 46, 16132–16137.

[51] M. Passare and H. Rullgård, *Amoebas, Monge-Ampère measures and triangulations of the Newton polytope.* Duke Math. J. **121** (2004), no. 3, 481–507.

[52] J.-P. Pin, *Tropical semirings.* Idempotency (Bristol, 1994), Publ. Newton Inst., 11, Cambridge Univ. Press, Cambridge, 1998, pp. 50–69.

[53] J.-P. Quadrat, *Max-plus algebra and applications to system theory and optimal control.* Chatterji, S. D. (ed.), Proceedings of the international congress of mathematicians, ICM '94, August 3-11, 1994, Zrich, Switzerland. Vol. II. Basel: Birkhuser. 1511–1522 (1995).

[54] J. Richter-Gebert, B. Sturmfels, and T. Theobald, *First steps in tropical geometry.* Idempotent mathematics and mathematical physics, Contemp. Math., 377, Amer. Math. Soc., Providence, RI, 2005, pp. 289–317.

[55] E. Shustin, *Critical points of real polynomials, subdivisions of Newton polyhedra and topology of real algebraic hypersurfaces.* Amer. Math. Soc. Transl. (2) **173** (1996), 203–223.

[56] E. Shustin, *Gluing of singular and critical points.* Topology **37** (1998), no. 1, 195–217.

[57] E. Shustin, *Lower deformations of isolated hypersurface singularities.* St. Petersburg Math. J. **11** (2000), no. 5, 883–908.

[58] E. Shustin, *Patchworking singular algebraic curves, non-Archimedean amoebas and enumerative geometry.* Preprint arXiv: math.AG/0211278.

[59] E. Shustin, *A tropical approach to enumerative geometry.* Algebra i Analiz **17** (2005), no. 2, 170–214 (English translation: St. Petersburg Math. J. **17** (2006), 343–375).

[60] E. Shustin, *A tropical calculation of the Welschinger invariants of real toric Del Pezzo surfaces.* J. Algebraic Geom. **15** (2006), no. 2, 285–322.

[61] E. Shustin, *Welschinger invariants of toric Del Pezzo surfaces with non-standard real structures*, Proc. Steklov Institute Math. **258** (2007), 219–247.

[62] D. Speyer and B. Sturmfels, *The tropical Grassmannian.* Adv. Geom. **4** (2004), 389–411.

[63] B. Sturmfels, *Viro's theorem for complete intersections.* Ann. Scuola Norm. Sup. Pisa Cl. Sci. (4) **21** (1994), no. 3, 377–386.

[64] B. Sturmfels, *Solving systems of polynomial equations.* CBMS Regional Conference Series in Mathematics. AMS, Providence, RI 2002.

[65] B. Teissier, *Amibes non archimédiennes.* In: Géométrie tropicale X-UPS 2008 (ed: P. Harnick, A. Plagne, and C. Sabbah), Ecole Polytechnique, 2008.

[66] W. P. Thurston, *Three-dimensional geometry and topology. Vol. 1.* Princeton Mathematical Series, 35. Princeton Univ. Press, Princeton, NJ, 1997.

[67] O. Ya. Viro, *Gluing of algebraic hypersurfaces, smoothing of singularities and construction of curves.* Proc. Leningrad Int. Topological Conf., Leningrad, Aug. 1982, Leningrad, Nauka, 1983, pp. 149–197 (Russian).

[68] O. Ya. Viro, *Gluing of plane real algebraic curves and construction of curves of degrees 6 and 7.* Lect. Notes Math. **1060**, Springer, Berlin, 1984, pp. 187–200.

[69] O. Ya. Viro, *Progress in the topology of real algebraic varieties over the last six years.* Rus. Math. Surv. **41** (1986), no. 3, 55–82.

[70] O. Viro, *Real algebraic plane curves: constructions with controlled topology.* Leningrad Math. J. **1** (1990), 1059–1134.

[71] O. Viro. *Patchworking Real Algebraic Varieties.* Preprint, 1995. Available at http://www.math.uu.se/ oleg/preprints.html.

[72] O. Viro, *Dequantization of real algebraic geometry on logarithmic paper.* European Congress of Mathematics, Vol. I (Barcelona, 2000), Progr. Math., 201, Birkhäuser, Basel, 2001, pp. 135–146.

[73] J.-Y. Welschinger, *Invariants of real rational symplectic 4-manifolds and lower bounds in real enumerative geometry.* C. R. Acad. Sci. Paris, Sr. I, **336** (2003), 341–344.

[74] J.-Y. Welschinger, *Invariants of real symplectic 4-manifolds and lower bounds in real enumerative geometry.* Invent. Math. **162** (2005), no. 1, 195–234.

[75] J.-Y. Welschinger, *Spinor states of real rational curves in real algebraic convex 3-manifolds and enumerative invariants*, Duke Math. J. **127** (2005), no. 1, 89–121.